ISSUES IN RADICAL SCIENCE

Radical Science 17
Edited by the Radical Science Collective

'. . . *an association in which the free development of each
is the condition of the free development of all*'

Free Association Books / London / 1985

Issues in Radical Science
Radical Science 17

Published in June 1985 by
Free Association Books
26 Freegrove Road
London N7

British Library Cataloguing in Publication Data

Issues in radical science.
   Radical science,
   ISSN 0305–0963; no. 17)
   1. Science—Social aspects
   I. Radical Science Collective   II. Series
   306′.45   Q175.5

   ISBN 0-946960-19-4

Cover illustration from Fritz Lang's *Metropolis*,
reproduced by courtesy of the British Film Institute

Cover design by Carlos Sapochnik
Typeset by Folio Photosetting, Bristol
Printed in Great Britain by SRP Ltd., Exeter

CONTENTS

FRITZ LANG'S *METROPOLIS:* Science, Machines and Gender

L. J. Jordanova

When *Metropolis* received its much-publicized Berlin première early in 1927, the critics and public alike were hostile to it (Kaplan). Many commentators found the ending of the film banal and unsatisfying, while they generally praised the modern images of machines and buildings which characterize the film and continue to be a source of admiration. *Metropolis* is a complicated and confused film. It draws, however, on some important themes relating to work, industrial organization and the nature of science which are particularly characteristic of European thought in the 1920s. Furthermore, it puts these themes into play through a plot which hinges on the nature of femininity — especially its twin aspects, virginity and sexuality — and on the role of women as the social and political bedrock of stable societies. This association of gender with analyses of science and technology was not novel; it drew on long-standing traditions which linked women with passion and superstition, and men with reason and knowledge (MacCormack and Strathern, eds.).

The story of *Metropolis* concerns a city-cum-industry run by Joh Fredersen in which the workers are reduced to a faceless mass of exploited bodies. A young woman, Maria, comforts them with reassurances that a saviour and mediator will come to deliver them from their anguish. Freder, the boss's son, sees Maria, falls in love with her and casts himself in the role of the people's deliverer and critic of his father. His father, however, learns that discontent is spreading among the workers and decides to enlist the help of Rotwang, the inventor, who has been working on a robot 'in the image of man, that never tires or makes a mistake' (Metropolis, p. 47 T, i.e. title in the film). Fredersen discovers that the workers are meeting secretly in the old catacombs to hear Maria talk to them about prayer and patience. After taking Fredersen to the catacombs to see Maria in action for himself, Rotwang captures Maria, imprisons her in his house and makes, at the boss's request, a robot in her exact likeness. The robot is made to incite the workers to revolt because Fredersen is looking for an excuse to use violence against them. Their uprising wreaks havoc and has the effect of flooding the underground city where the workers live, thus putting in jeopardy the lives of their children. The workers, thinking she has destroyed their children, pursue Maria, but capture the robot and burn it as a witch, thereby revealing its true nature — a machine, not a person. Rotwang chases the real Maria onto the roof of the cathedral from where he falls to his death after Freder goes to her rescue and fights him off. Maria, father and son are reconciled and a workers' leader comes forward in the same spirit; for it was, after all, Fredersen's son who had saved their children. The boss is symbolically united to the workers by a handshake at the end of the film.

To understand the film, we need to know something of its conditions of production. Fritz Lang (1890–1976) was born in Austria and had trained as an architect and artist before turning to the film industry, in which he worked as an actor and scriptwriter, coming to prominence as a director in the 1920s. Lang thought of setting a film in a futuristic city during a visit to New York in 1924. His wife and close collaborator, Thea von Harbou, then wrote a novel upon which the film, in turn, was based. The original film, shown in Germany, was longer than the version currently available and apparently contained characters and events from the novel which are now missing. The version we know was made for the United States and has been considerably altered. No copies of the original are known to exist. To speak of the film as a production of Lang's thus constitutes a significant simplification; it was the work of many hands, but it is impossible to know exactly what the terms of the collaboration were or what the effect of the cuts was. Furthermore, in later years, as an exile in the United States, Lang was quick to criticize *Metropolis* and its romantic, simplistic ending. Although all these points lead to interpretative problems, they do not undermine the possibility of a historical analysis of *Metropolis*, which needn't depend on Lang's special status as the main creator of the film.

It may be useful, however, to note the main respects in which the novel and the film (as it is known) diverge. In the novel, Fredersen and Rotwang are locked in mutual hatred over their love for Hel, Freder's mother who had died when her son was born. Fredersen 'stole' Hel from Rotwang. Von Harbou situated Rotwang in an ancient magical tradition by explaining the uniqueness of his house — a medieval island in a sea of skyscrapers — in terms of an earlier occupant who had possessed awesome powers. Similarly, she accounts for the anomalous survival of a Gothic cathedral through the power of the group of monks who still run it. Furthermore, Fredersen has a mother from whom he is estranged because she disapproves of his general conduct. His reconciliation with her concludes the book and carries with it a pledge that *he* will reform, rebuild and redeem Metropolis. In the novel, Rotwang's death results from his belief that Maria, whom he sees in the cathedral, is his beloved Hel; he cannot understand why she flees from him — a mistake which is comprehensible only in the context of a fight he has had earlier with Fredersen. When he regains consciousness following this, he believes himself dead, and so goes in search of his lost love. At the level of the plot, therefore, the novel is fuller and more consistent than the film and contains significantly different emphases. The use of florid religious imagery is much more elaborate, the references to father/son conflict more overt, and the symbolism generally more highly developed. I would guess that the differences stem not only from the cuts referred to above, but also from the translation of verbal into visual images.

Weimar Tensions

The difficulties in interpreting the film are of two main kinds. The first stem from the peculiar historical circumstances of pre-Nazi Germany, the use of films as instruments of Nazi propaganda and the attempt to come to terms with fascism following World War Two. This issue is often reduced to a concern with the question: Was Lang in general, and his work in *Metropolis* in particular, marked by the same ideological tendencies which led to the rise of fascism? In other words, is it necessary to find ways of dismissing them as morally and politically tainted? This drive for moral clarity has led, for example, to a debate about whether Rotwang is a precursor of the reviled Jewish figures of later Nazi films — a point to which we shall return later. The 'problem' of Lang has been solved in a number of ways, one of which is to attribute blame for any apparently unsound ideological tendencies in his films to the contributions of Thea von Harbou, who remained in Germany after Lang left, and was an active film-maker under Hitler. Yet, to pose the question of Lang's political views in this way is to make assumptions about the second interpretative issue — the relationship between cultural products such as film and the social setting in which they are made. Theoretically this is a particularly hard issue to deal with, and few attempts have been made to do so.

To simplify, the issue is whether Lang is merely reflecting general, even unconscious tendencies in his own culture, which are the very ones that made Hitler's rise to power possible (Kracauer), or whether he is putting forward the views of a specific group with a coherent ideological perspective (Tulloch). Significantly, another possibility, that Lang's is a highly idiosyncratic vision, is little entertained because critics generally wish either to exonerate him from or implicate him in broader social movements of the 1920s. It is, I think, unsatisfactory to see him either as a passive reflector or as the mouthpiece of a particular group, for reasons which will be explored later in connection with the specific themes of the film. For the moment a more general difficulty should be borne in mind. If we take it for granted that cultural artefacts are in some sense socially produced, then we need to search out and lay bare the various levels of mediation between economy, society and culture. In the case of Weimar Germany this is an intricate process; no literature exists which seeks to carry out such a job.

Of course, the standard cultural histories of the period make many assertions about these relationships, based on various theoretical suppositions and prejudices, but they fail to work out the links in any systematic way. For example, in his highly acclaimed work, *Weimar Culture*, Peter Gay locates *Metropolis* among works which portray 'the revenge of the father'. He finds it a film 'calculated mainly to sow confusion', a 'tasteless extravaganza' and 'a repulsive film' (pp. 148–49). He concludes his account of *Metropolis*: 'The revenge of the father and the omnipotence of the mother were twin aspects of the Weimar scene, both equally destructive to youth' (p. 149).

Such an approach clearly cannot shed light on the highly specific fashion in which the film portrays the workplace and the labour process. We can juxtapose this portrayal with what is known about labour conditions, wage settlements and the introduction of industrial rationalization in the period. The links between these two levels, the material conditions and their representation, need to be systematically examined. I have, however, been unable to locate any rigorous attempts to look at how labour was represented in a variety of cultural settings (art, film, theatre, fiction, social theory) and to offer an overall interpretation of the way labour/capital relations were treated. What is said about this later on must therefore remain speculative.

The difficulties of interpreting a film which is produced in such a fraught context mean that we must be especially careful about attributing a moral position to its director. This is in part because such positions are rarely articulated unambiguously, and also because it is hard to know how the ordinary public understood the film at the time it was produced. The opinions of critics, while illuminating, are not necessarily representative. And, if we want to make assertions about *Metropolis* as an expression of the conscious and unconscious tendencies of its time, it helps to have some independent means of assessing what these tendencies were. My point is that these are frequently inferred by hindsight, starting from the subsequent ascendancy of fascism. This teleological approach is understandable, since our need to distance and purify ourselves from the Nazis is still very strong, as the persistent popularity of films about the Second World War containing stereotyped Germans testifies; yet it is also unhelpful. It is therefore necessary to situate *Metropolis* in its historical context.

Cultural histories of Weimar which mention *Metropolis* generally present it in terms of crises of belief and identity, highlighting the religious and Oedipal themes (e.g. Gay and Laqueur). The film certainly does explore a number of easily recognizable Christian themes: Maria, the Virgin-mother; a son striving to save the world; a stern, almighty father; the virtues of patience and prayer; the necessity for suffering in order to overcome evil. These themes are even more heavily underscored in the novel, in which Fredersen is locked in conflict with the monks of the Gothic cathedral who believe that doomsday has come when the city is in turmoil. Furthermore, Fredersen himself experiences the cataclysm as an occasion for repentance and he seeks to become the new redeemer of Metropolis. In the film, the use of crosses in the catacombs where Maria comforts the workers, of a halo of light around her head, of the Tower of Babel parable and even the frequent use of triangular motifs (the Trinity) further reveal an indebtedness to traditional religious language. Equally evident is the Oedipal theme. Freder rebels against and wishes to destroy his father. Indeed, in her novel Thea von Harbou talks explicitly of his parricidal drive. Historians have found the conflict between father and son revealing of the general cultural crisis of Weimar. Hence, the ending of the film — in which father and son are reconciled, yet without any radical change in the power structure being on the cards — appears especially prescient of the rise of totalitarian power (e.g. Kracauer).

The treatment of religion and inter-generational conflict — like that of other motifs in the film such as the double, Frankenstein and Faust — will be best understood through a systematic comparison of *Metropolis* (both the book and the film versions) with other cultural productions of the period — films, plays, poems, theatre, ballet, the visual arts, architecture and music. Unfortunately such a project has not yet been undertaken.

My concern here is with the deployment of science and technology within the film, and more particularly with the ways these are related to magic and tradition on the one hand, and the dual nature of female sexuality on the other. Those who have emphasized science and technology often classify *Metropolis* as 'science fiction': 'a class of prose narrative which assumes an imaginary technological or scientific advance, or depends upon an imaginary and spectacular change in the human environment' (*Oxford Companion to English Literature*, p. 734). At first sight the use of the robot further reinforces the status of *Metropolis* as science fiction. Yet *Metropolis* was conceived as an expression, if a somewhat exaggerated one, of a city life already firmly rooted in American culture. The robot, in the sense of an artificially made human being, relates as much to ancient myth as it does to a projected future, and Lang's film, like von Harbou's novel, is striking for the persistence of historical reference. The clothes are not futuristic but contemporary or traditional, the language and value systems are those of the 1920s and its parent culture, the modes of transport those in common use. Even the machines, which might possibly evoke an idea of 'technological or scientific advance', exist more as primitive deities than as modern marvels. In short, to categorize *Metropolis* as science fiction draws our attention away from its use of science and technology *in dynamic interplay with* magic and tradition. The film lays bare the exceedingly fragile boundaries between good and bad science, good and bad beliefs, good and bad machines, good and bad women.

The current vogue for *Metropolis* — as I write, a new tinted version with a pop sound track is on release — is undoubtedly connected with the conventional view of it as sci fi. The robot image, which many associate with the film, pervades much contemporary popular culture: it is a fun symbol with a touch of kitsch. Yet, at the same time, talk of the degradation of work, particularly as a result of the silicon chip revolution, is now a commonplace — rekindling the fear of human workers becoming reduced to mere 'hands' servicing tyrannical machines. Reactions to *Metropolis* in the 1980s are thus, I suspect, as ambivalent as they ever were — composed of an aesthetic appreciation of its unrepentant modern imagery, on the one hand, and a revulsion against its human implications on the other.

Workers or Slaves?

Here I wish to emphasize four topics: industry, science and technology, city life and modernism. *Metropolis* is set in a city which is also a single industrial plant with one man in charge of everything. The workers service the machines, which require constant attention; thus, while both human labour and mechanical power are required to keep the Metropolis going, the former are subservient to the latter. The worker must keep up with the machine, and this is unambiguously shown as the source of excessive fatigue over long shifts.

Two themes prominent in contemporary debates about industrial organization are evoked in *Metropolis* — scientific management in its broadest sense, and the role of corporations. As a movement, scientific management is commonly linked with the American engineer Frederick Winslow Taylor (1856–1915), whose work became widely known in America in 1910 as a result of a government inquiry and whose *Principles of Scientific Management* (1911) had been translated into German in 1913. Taylorism built on earlier moves towards 'systematic management' which had stressed the importance of a system of management for directing and controlling production (Litterer, Nelson). Such streamlining of administration, through centralizing and standardizing managerial tasks to avoid wasted effort, is forcefully expressed in the depiction of Fredersen's office. As developed by Taylor, the theory of scientific management was strongly committed to rationality and efficiency. It also entailed finding the best person for each task, breaking down jobs into their constituent tasks in order to analyse how each one could be undertaken in the most efficient manner and then training the workmen to use this (and only this) approach. Taylor and his followers maintained that this would dramatically increase efficiency and so productivity. Something of the flavour of his system can be gleaned from his remark that '[the] work [of handling pig iron] is so crude and elementary in its nature that the writer firmly believes that it would be possible to train an intelligent gorilla so as to become a more efficient pig iron handler than any man can be' (Taylor, 1911, p. 40).

The implication — occasionally made explicit by Taylor, that less 'human' men make better workers — is clearly taken up in *Metropolis*, where the labourers move in a senseless mass, devoid of individuality. Lang also showed the workers as dominated by time. The shifts in Metropolis last ten hours, and the clocks appropriately have a ten-hour face. Not only was working to fixed time schedules central to early industrialization (Thompson), but scientific management extended this through the emphasis on the timed task, the importance of avoiding wasted effort, the need for production schedules and the setting of wage rates and bonus systems.

Metropolis thus captured some aspects of scientific management — the subservience of people to the work process and the tyranny of time — and exaggerated them, as in a caricature to heighten the viewer's sense of industrial inhumanity. Other features of scientific management, however, find no expression in *Metropolis*. Two silences in particular stand out. First, the role of management and of technical expertise was central to Taylorism, which was unthinkable without both the enthusiastic cooperation of the managerial strata and the expertise of engineers. These groups of middle-class professionals are never seen in the film, yet in the social vision of scientific management they played a crucial part, for reasons which will become clear when the second silence has been identified. This concerns rewards for work. Taylor and his followers believed that fair wages were of the utmost importance and that higher productivity was directly in the workers' interest, because it would lead to higher wages. The reasoning behind this was perfectly plain — higher incomes undermined class solidarity, enhanced social mobility, and, through the power to consume which better incomes offered, drew working people into a middle-class life style. In theory at least, the lure of moving into the professional and managerial classes would undermine any possible discontents. Of course, the discourse of scientific management is not free from tensions and inconsistencies. The goal of a classless, stable society fits ill with the emphasis on the intense specialization of work which Taylorism requires. Significantly, *Metropolis* portrays work as physically demanding rather than needing specialized skills, while the workers are never shown making products or having or spending money; in stark contrast to the rich, the workers are exhausted, walk like zombies, get killed in industrial accidents, work in hot, steamy conditions and thus lead miserable lives. In this respect their status resembles that of slaves, not that of the modern workers scientific management sought to create.

During the 1920s there was intense concern with the growth of large industrial complexes and monopolies as these assumed ever greater political and economic power. Metropolis was just such a body — a single giant unit, city-state and factory rolled into one. Corporatism constituted an ideology rooted in the transfer of power away 'from elected representatives or a career bureaucracy to the major organised forces of European society and economy'

(Maier, 1975, p. 9). Maier associated this shift with a weakening of parliamentary democracy, the growth of private power and an erosion of the distinction between the public and private sectors, and the development of centralized bargaining procedures in which labour leaders played a significant role. He found Germany moving clearly towards corporatism during the 1920s, a trend he identified with conservatism.

Aspects of these themes are certainly explored by Lang. In *Metropolis* there is no political structure in which people can participate, hence any distinction between public and private is totally inapplicable. If we understand private power as suggesting both the ascendancy of particular interests and the dominance of individuals, then Fredersen represents such power. On the other hand, there is no hint of bargaining between major groups in the film, not least because labour has no voice, being reduced to a collection of faceless bodies. The ending may be thought to hold out the promise of a negotiated settlement, but this can be no more than conjecture, since its emphasis is a sentimental rather than a practical one. Furthermore, the workers are portrayed in distinctly unflattering terms: they are unable to distinguish the false from the true Maria; they can easily be roused to violence which is injurious to their own families; they lust after revenge against (the robot) Maria; and they become instantly docile once their children are known to be safe.

It could be argued that such a portrait merely served to highlight the degradation of the system in Metropolis and so sharpen a critique of modern industrialism. It seems more likely, however, that Lang took up certain themes which appealed to him dramatically, and then developed them in an extreme form without feeling bound to produce a logically consistent whole. What *Metropolis* reveals about industry, then, is not the historical location of corporatism and scientific management in Weimar Germany, but an unfocused nexus of tensions and problems which industrial development was seen to be spawning, concerning particularly the relationships between workers and machines and the reduced human potential of industrial workers.

Clearly, it would have been impossible for these themes to be taken up in *Metropolis* without science and technology occupying a visible public position. The film treats science and technology in two highly specific ways, both of which owe less to contemporary events than to long-established literary and artistic themes. We have already mentioned how the machines the workers tend also dominate and control them. Apparently machines keep Metropolis going, though exactly how they do so remains unspecified. Lang can thus draw on a naive faith in technology and, simultaneously, express a primitive fear of machines when he turns them into monsters, named after non-Christian deities, who swallow up workers just as primitive gods demanded the constant sacrifice of human victims. Either way the machines are rendered omnipotent — either because of moden technology or because of irrational belief. Such ambivalence about machines was in no way specific to the 1920s, although it may have been fuelled by contemporary industrial developments.

Metropolis also draws on the old tradition concerning the artificial production of human beings. It is true that this is achieved not directly from organic remains, as in *Frankenstein*, but via a robot and elaborate machinery. From the imagery of the dissecting room and charnel house we have moved to that of the modern physics and chemistry laboratory. When Rotwang the inventor creates the false Maria, it is flashing lights, flasks and electrical phenomena which we see. More revealing still is the fact that Rotwang has created the robot in the first place as 'a machine in the image of man, that never tires or makes a mistake' (*Metropolis*, p. 47, T) in an evident allusion to automated production; 'the workers of the future — the machine men!' Yet Rotwang looks not like a modern scientist but like a hermit who knows about magic and alchemy, wears a long gown, has a somewhat demented manner, old books, and a medieval house nestled incongruously in the modern city and harbouring mysterious secrets. Rotwang is a complex figure who displays prodigious intellectual powers which even Fredersen must respect and which are of infinite value in the modern Metropolis, and who also manifests the archaic powers of a sorcerer. In this respect he resembles the machines, at once suggestive of modern power and primitive evil.

The City

If early twentieth-century thinkers wanted to voice reservations about the times in which they lived, the city offered an attractive vehicle for their doubts. The city could stand for a multitude of discontents, as it had done for centuries. Yet, for equally long, the city had also represented positive values such as learning, civilization and enlightenment. In *Metropolis*, Lang highlights modern, high-rise architecture, advanced transport systems and the vertical structure of the city as a representation of its social hierarchy. Lang had, himself, trained as an architect, the profession of his father, although he showed no enthusiasm for the job. He was particularly keen in *Metropolis* to find novel cinematic ways of conveying the immense height of the buildings. What contemporary commentators found troublesome about city life, particularly in the United States, was the close proximity between different social, religious and ethnic groups. It was perceived as a location which threatened communities, for these could hardly hold together amidst the insistent mobility of urban life. Some who have written about Lang have pointed to distrust of the city as a characterstic of German conservative thought in the 1920s. Yet contemporaries found other meanings in modern city life apart from the threat of social disintegration; these often indicate a kind of exhilaration which went with being free from the oppressive intimacy of rural or small town life (Simmel). Lang shows the Metropolis not as a place where groups mingle promiscuously but as one place where they are rigidly segregated. The public places are only for an elite whose composition is never specified and whose behaviour vividly evokes the decadent pleasure-seeking with which the Weimar period is so often associated.

However, there can be no doubt that the city stood for modern life, or that modernity was an important feature of Lang's film. The modernity is conveyed in a number of ways, which do not always sit easily together. The self-indulgent merry-making of the privileged elite is one way of suggesting it; others are the modern architecture, the industrial machinery, the transport systems and Fredersen's bare, functional office. These must be seen, however, in relation to the ancient catacombs, Rotwang's medieval house, the Gothic cathedral and the eighteenth-century costumes in which the gilded youth of Metropolis besport themselves. Thus the film does not present a simple futuristic or modernistic scenario, but sets up a dynamic between new and old. It is worth remembering how very controversial, socially and politically, simple functionalist architecture was at this time, and that the underlying issue was not stylistic preference but an entire worldview.

In so far as *Metropolis* rests on a worldview, it is organicism and not modernism which provides a unifying theme. Frequent allusions are made to the need for an integrated harmony between different parts of the body: 'Between the brain that plans and the hands that build, there must be a mediator'; 'It is the heart that must bring about an understanding between them' (Metropolis, p. 60 T). The film in fact ends with the following title: 'There can be no understanding between the hands and the brain unless the heart acts as mediator' (p. 130). These statements imply that a society must function as a whole system, just as an organism does;

that social unity is a prized value. Yet this is put forward in a context of intense exploitation and extreme division of labour, a fragmentation which the organicist formulation does not seek to challenge. The concern with bringing together head and hands indicates a deep fear of splitting in the social order which will be mended *not* by ending the original divisions, but by binding groups together in some unspecified way through the language of emotions and sentiment. The organicist discourse in *Metropolis* works, then, at two levels: the first stresses harmonious relations within a social system, while the second registers divisions and hierarchy. The point about the hierarchy, however, is that each stratum depends on the others and therefore has no autonomous existence.

Visually, the distinction between the different levels is powerfully conveyed. The workers wear sombre clothes and live below ground where it is dark, while the elite live high up, travel in airplanes, wear pale clothes and experience open air. The brain, the organ of calculation and hard thinking, is visually expressed in silent-film style by the exaggerated emotions and facial contortions of Fredersen and Freder. The workers, on the other hand, are purely physical; they are 'hands', and hands in another sense when they are shown moving dials on clocklike machines, their arms like the hands of a clock. Not only does the life of the mind not exist for them, but they are barely differentiated from one another. The balletic presentation of work, which shows highly abstract movements, heightens this sense that mechanical coordination between identical elements represents the sum total of life for the majority of the inhabitants of Metropolis who trudge to and from work in serried ranks. The congruence between organicism and silent film technique extends to Maria, who, as the 'heart' in the system, constantly presses her hands to her breast.

The workers appear to be all male, so that for the most part Maria is the only woman we see. The women workers/workers' wives become visible only when the masses rebel and then become alarmed about the fate of their children. The whole plot in fact rests on the potentially disruptive presence of Maria. Her good, pure femininity — she is both Virgin and mother (Metropolis, p. 27 N, i.e. a quotation from the novel) — is an essential part of the organicist vision, for it enables her to be the 'heart' of the city. It is equally important that the robot be her double — outwardly identical but inwardly her opposite. Femininity is thereby split into two: pure, good chastity and sensual, corrupt depravity. Gender comes to play a complex role in the film. The real business of life, whether it is labour or running Metropolis, is done by men, yet they lack some essential element to make them whole, and it is this ingredient which good femininity can contribute. So that although reason and sentiment could be seen as being in opposition to one another, they are also complementary.

But what of the destructive side of femininity? Full sensuality is presented as a form of unreason closely akin to mass fury and mass decadence which cannot be fused with male reason. Its fate is to be identified as witchcraft and suitably annihilated. The robot, built by Rotwang, a master of knowledge, is portrayed as the antithesis of such knowledge. (It should be pointed out that the paradox is sharpened by the robot looking unmistakably feminine even before it is turned into 'Maria', suggesting that the destructive machine and the destructive side of female sexuality are closely identified with one another.) Some commentators have dealt with this problem by suggesting that Rotwang is the black magician, the precursor of the evil Jew of Nazi films. Equally he could be seen in the tradition of hermits, alchemists, philosophers and anatomists, shown in paintings as old men wearing robes, whose knowledge isolates them from others. The pentagram, shown on his front door and in his laboratory, is, when inverted, a sign of witchcraft and inverted human nature. The term 'seal of Solomon' is also used, but this is in fact six-pointed, unlike the shapes shown in the film.

To identify Rotwang with bad magic is to say that the power he has is wholly illegitimate and comes from his attempts to do things which are beyond the proper province of the human being. There is certainly some truth in this, and it accounts for the connections often perceived between *Metropolis* and *Frankenstein*. Yet, his power is genuine in the sense that it really works, he has a level of understanding which, it is implied, is unique and valued. Rotwang cannot simply be a magician, then; he is also a scientist and inventor who grapples with the real world, not the realms of fantasy. Indeed, if he were simply a wizard, the film would be far less interesting, not least because Fredersen's invitation to Rotwang to help him solve his problems would seem less plausible. The alliance between political power and power over nature appears forceful indeed, while it also serves to show that legitimate power and knowledge can all too easily enter an unacceptable domain. The boundary between good and bad here appears perilously fragile.

Power to Reconcile

Metropolis is an exploration of pure power. Fredersen has complete authority and control, just as Rotwang can command nature's forces. Yet both men find their power challenged, because, the film implies, it was incomplete psychologically. The plot resolves this by eliminating Rotwang (for reasons which remain unclear in the film) and by giving Fredersen the capacity to empathize. His power — political, social, economic, and technological — remains intact, but something is added to it to make it whole. This something is unambiguously identified as a feminine virtue, though some men, such as Freder, can learn it. The need for good femininity to fill the lacuna in male power is reinforced by the fact that Freder has no mother to mediate between him and his father, so Maria has to fulfil this role. She does this not directly but indirectly by creating tenderness in the son, who then tries to pass it on to his father in order to bring his father to sympathize with the workers. The film thus works with a number of different kinds of power and the relationships between them: the power of the emotions (Maria and Freder), of the capacity to control nature (Rotwang), of absolute political authority (Fredersen), of wanton destruction (the robot, the masses and the monstrous machines).

Metropolis reveals much about the relationship between science and technology and other forms of power; or rather, science and technology offer Lang a verbal and visual language with which to speak about social relationships and political structure. The relationship between mental and manual labour is identical with that between rulers and ruled; the unifying force is the highly mechanized factory which is also the state — an organic social system. The visual image is that of vertical hierarchy; the verbal image that of physiological systems. Femininity, triggering sexual attraction, is the dynamic element which introduces change into the system. And, in the end, only Maria can offer the quality, heart, which will reconcile head and hand. If the feminine disrupts, it also heals. Women do not represent a unitary power but a force which easily fragments into opposites like chastity and sensuality.

In a similar fashion, scientific knowledge can split into genuine reason and illegitimate knowledge/magic. Technology also potentially contains the all-powerful, efficient machines and the monsters who claim the lives of men. When concepts split in this way, they generate tensions — both because the relationship between the two elements may be obscure and troublesome, and because each element can easily be transformed into the other and hence create instability. Such unstable splits threaten to destroy the organic state; they require bridging, just as there should be links between labour and capital. The bridges do not undermine the divisions they span but rather provide an illusion of cohesion.

Lang's film is best described as a caricature of modern life which exaggerates certain aspects to bring them to our attention. This method is most successful in relation to the workers. The denial of individuality, the fusion of man and machine, workers going to and from work in mindless synchrony, and the obsession with time and efficiency have been noted in critiques of capitalism since the time of the romantic writer and historian, Thomas Carlyle (1795–1881). But, it remains an open question what Lang and his co-workers really believed about these features of modern life. The film seems to have a conservative, palliative ending, in that master and workers are reconciled without anything being said about real material improvement. The conclusion constitutes a romantic promise while the reality is that the organic system is preserved intact and of course remains hierarchical and exploitative. The use of organicist imagery, and the extended religious analogies surrounding Maria in particular, leave all the important political questions not raised, let alone answered.

It would be wrong, however, to allow the banality of the ending to colour our reactions to the entire film. Commentators are virtually unanimous in finding the striking visual effects Lang deployed a brilliant success. This is not, I think, to reduce discussion of the film to purely aesthetic terms but rather to acknowledge the source of its impact. Visual images play a crucial role in exploring and working with the themes discussed in this paper. They expose to our view some of the conflicts and complexities associated with modern industrial and political systems. Until the analysis of television in recent years, those of us concerned with the social and political impact of science have scarcely noticed how important visual images are. It is an inescapable conclusion that our understanding of science and technology will be enormously enhanced when we can read more of the codes, symbols and images which convey, often in historically specific ways, something of the power, authority and control knowledge of nature offers. These complex visual languages speak to our imagination and are all the more important because they do so, since they readily combine with taken-for-granted assumptions about such issues as gender, just as *Metropolis* does. Acknowledging that the power and nature of fantasies are fit subjects for study offers some important analytical tools with which science and technology can be understood in their cultural context.

Acknowledgements

This paper arose out of a course I taught in 1982 on 'Tradition and Modernity' with Steve Smith, to whom I am grateful both for a pleasurable teaching experience and for his subsequent help and encouragement. I would also like to thank Ann Barnard and Janet Bloomfield for bibliographical suggestions, the members of the Department of Art History and Theory Graduate Seminar, University of Essex, for their constructive responses to a version of this paper, and Karl Figlio for his generous help at all stages of the work. Without the assistance of the Inter-Library Loan Department, University of Essex, I could not have had access to so much valuable material.

References

All items published in London unless otherwise stated.

J.D. Barlow, *German Expressionist Film*, Boston, Twayne, 1982.

M. Berman, *All That Is Solid Melts Into Air. The Experience of Modernity*. Verso, 1983.

S.E. Bronner and D. Kellner, eds., *Passion and Rebellion: The Expressionist Heritage*, Croom Helm, 1983.

T. Carlyle, *Past and Present*, Chapman and Hall, 1843.

L.H. Eisner, 'The German Films of Fritz Lang: Some Impressions', *Penguin Film Review*, 6 (1948), 53–61.

L.H. Eisner, *The Haunted Screen: Expressionism in the German Cinema and the Influence of Max Reinhardt*, Secker and Warburg, 1973.

J. Elderfield, 'Metropolis', *Studio International*, *183* (1972), 196–199.

P. Gay, *Weimar Culture: The Outside as Insider*, Harmondsworth, Penguin, 1974.

T. von Harbou, *Metropolis* (1927), New York, Ace, 1963.

S. Jenkins, ed., *Fritz Lang: The Image and the Look*, British Film Institute, 1981.

P.M. Jenson, *The Cinema of Fritz Lang*, Zwemmer, and New York, A.S. Barnes, 1969.

P. Joannides, 'Aspects of Fritz Lang', *Cinema* (Cambridge), (August 1970), 5–9.

E.A. Kaplan, *Fritz Lang: A Guide to References and Resources*, Boston, G.K. Hall, 1981.

S. Kracauer, *From Caligari to Hitler: A Psychological History of the German Film*, Princeton, New Jersey, Princeton University Press, 1947.

F. Lang, 'Happily Ever After', *Penguin Film Review*, 5 (1948), 22–29.

F. Lang, *Metropolis*, Lorrimer, 1973 (cited in the text as Metropolis; a detailed description of the film, with quotations from the novel, and the text of the titles).

W. Laqueur, *Weimar: A Cultural History 1918–1933*, Weidenfeld and Nicolson, 1974.

J.A. Litterer, 'Systematic Management: The Search for Order and Integration', *Business History Review*, *35* (1961), 461–476.

R.D. MacCann and E.S. Perry, *The New Film Index: A Bibliography of Magazine Articles in English, 1930–1970*, New York, E.P. Dutton, 1975.

C. MacCormack and M. Strathern, eds., *Nature, Culture and Gender*, Cambridge, Cambridge University Press, 1980.

C.S. Maier, 'Between Taylorism and Technocracy: European Ideologies and the Vision of Industrial Productivity in the 1920s', *Journal of Contemporary History*, 5 (1970) 27–61.

C.S. Maier, *Recasting Bourgeois Europe: Stabilization in France, Germany and Italy in the Decade After World War I*, Princeton, New Jersey, Princeton University Press, 1975.

Metropolis — see under F. Lang.

P. Monaco, *Cinema and Society: France and Germany during the Twenties*, New York, Oxford, and Amsterdam, Elsevier, 1976.

D. Nelson, 'Scientific Management, Systematic Management, and Labor, 1880–1915', *Business History Review*, 48 (1974), 479–500.

Oxford Companion to English Literature (fourth, revised edition), Oxford, Oxford University Press, 1983.

H.S. Person, 'Scientific Management', *Encyclopedia of the Social Sciences*, 13–14 (1930), 603–608.

G.D. Phillips, 'Fritz Lang on Metropolis', in T.R. Atkins, ed., *Science Fiction Films*, New York, Simon and Schuster, 1976, pp. 19–27.

E. Rhode, *Tower of Babel: Speculations on the Cinema*, Weidenfeld and Nicolson, 1966.

M. Shelley, *Frankenstein, or the Modern Prometheus* (1818), Oxford University Press, 1969.

G. Simmel, 'The Metropolis and Mental Life', in D. Levine, ed., *Georg Simmel on Sociability and Social Forms* (1903), Chicago, University of Chicago Press, 1971, pp. 324–39.

F.W. Taylor, *The Principles of Scientific Management*, New York and London, Harper, 1911.

F.W. Taylor, *Die Grundsätze Wissenschaftlicher Betriebsführung*, München, R. Oldenbourg, 1919 (first German edition 1913).

E.P. Thompson, 'Time, Work Discipline and Industrial Capitalism', *Past and Present* 38 (1967), 56–97.

J. Tulloch, 'Genetic Structuralism and the Cinema: A Look at Fritz Lang's Metropolis', *Australian Journal of Screen Theory* 1 (1976), 3–50.

J. Willett, *The New Sobriety 1917–33: Art and Politics in the Weimar Period*, Thames and Hudson, 1978.

A. Williams, 'Structures of Narrativity in Fritz Lang's *Metropolis*', *Film Quarterly*, 27 (1974) 17–24.

R. Williams, *Keywords: A Vocabulary of Culture and Society* (revised and expanded edition), Fontana, 1983.

BRITAIN'S MINERS AND NEW TECHNOLOGY

Dave Feickert

What has the 1984–85 miners' strike got to do with new technology? A great deal. Indeed, the government's plan for mass redundancies follows inexorably from automated equipment installed in recent years and from the new high-tech capacity to be installed over the next few years. This brief article will sketch the origins of today's battles in the miners' 1974 victory, particularly in the resulting 'Plan for Coal', which has come to be manipulated by capital against labour.

For all of us 1974 was a crucial year. It was one of those turning points around which history revolves. Although we did not realize it at the time, it was effectively the last year of the long postwar economic boom. But as such it was also the door opening into a quite new period of profound restructuring of the international economy. One of the levers of this restructuring would quickly become apparent — information technology, based on the revolution in microelectronics.

The year 1974 was also the moment when the long wave of post-war working-class struggles reached its highest point. In Britain this was revealed in the 1974 miners' strike, the three-day week and the fall of Edward Heath's Conservative Government. Parallel struggles had been fought out in all other developed capitalist countries as the international cycle of struggle circulated. The heart of the fight was the wage offensive that caught up almost all major groups of workers in those countries.

At the same time Third World movements were experiencing reinvigoration, perhaps mostly seen in the struggle of the Vietnamese people, already near victory, against the technological might of the most powerful nation on earth. In the Middle East the dispossessed Palestinians were at the centre of an unfolding political process which launched not only the Arab-Israeli war but also an exceptional fight for the redistribution of wealth, oil wealth, to the Arab producer nations of the Middle East. In 1973 they unleashed the world oil crisis that then began rocketing through the world economy in 1974, thereby providing the incentive for the massive restructuring which we now witness.

The multinational oil companies, together with local Arab elites, surfed in on this huge wave of underemployed and peasant struggle, yet the struggle itself found a material, if not political, unity in Britain. This was no more clearly seen than during the 1974 miners' strike when Welsh miners demonstrating for the strike outside the National Union of Mineworkers HQ in London dressed themselves in Arab clothing.

'Plan for Coal' 1974

As these two powerful struggles intermingled, the new crisis became particularly acute for Britain. Oil prices were going through the roof, which for an energy-intensive nation was a terrifying experience. At the same time it became all too apparent that the only viable alternative to oil — British coal — had been run down to an almost disastrous degree. Miners' leaders had been warning of the dangers throughout the period of rundown, to both Conservative and Labour governments alike, but to no avail.

It was in this context that the long-standing 'Plan for Coal' was pulled out of a drawer at Hobart House, HQ of the National Coal Board (NCB), dusted off and laid on the table of a specially set up Tripartite group of government, employers and trade unions. Although some of the proposals had been sitting on various tables for some time, the final text of the 1974 Plan was written with the recent triumph of the miners' union behind the pen.

The new version of 'Plan for Coal' became the second victory scored by miners within the same year, the first being over wages. But like so many working-class victories, the Plan contained several elements that were open to manipulation and recuperation. Ten years later, the many hours of negotiations between the NUM and the NCB have arisen from this problem. The Plan, together with its 1977 update, agreed on four aims:

* to modernize the coal industry, investing substantial sums where neglect once ruled;
* to replace exhausting pits with brand new ones, like Selby in North Yorkshire, and bring in 42 million tonnes of new capacity;
* to expand the market for coal to 135 million tonnes at least by 1985;

* to bring the industry's finances up to date, taking into account the social costs of the past.

During negotiations between the two sides in 1984, the argument revolved around the interpretation of 'Plan for Coal'. That, at least, was the public projection of it. The NCB insisted on negotiating on what it called the 'principles of Plan for Coal', while the NUM insisted on holding to the 'Plan for Coal' itself. In its briefing package to the TUC, the NCB went some way towards outlining the principles of the Plan as it saw them. Of the above four points, it entirely left out expansion of the market and vigorously interpreted clauses relating to the closure of exhausting capacity as a carte blanche to close pits on economic grounds, even if they still had coal reserves.

The Coal Board felt justified in thereby screening out of existence one of the Plan's two most important principles — expansion of the market — because the energy market as a whole had contracted sharply since 1973, particularly after 1979. On the other hand they pointed out that they had made the promised commitment to invest substantial sums in the industry and to modernize it. All the new capacity, agreed under the Plan, was being introduced.

Investment and Restructuring

While the NCB took few new marketing initiatives in the face of the declining energy market, they maintained high levels of investment both in new capacity and in reconstructing older coalfields, albeit all within the central coalfield. In this way it can be seen that the NCB have played off the growing contradiction between falling demand and an ever-increasingly productive supply of coal. Miners' jobs, caught in this vice, have been squeezed out in their thousands.

Moreover, the Plan for Coal had envisaged the maintenance of mining employment, as miners transferred from the old exhausting pits to the new ones or retired. In retrospect this assumption was misplaced, though understandably at the time. What was not realized at the time, especially by miners' leaders, was that the investment they had demanded for so many years would itself be used to cut thousands of jobs.

'Plan for Coal' was drafted in 1974 and redrafted in 1977. In 1974 the world's first commercially available microchip had only just appeared, and while engineers at the NCB's Mining Research and Development Establishment (MRDE) will have understood the potential use of microelectronics at that time, no one else did. Since then, however, investment in microelectronics has been used to transform mining operations both underground and on the surface, in collieries, offices and workshops, *in the central coalfield*.

Miners' leaders assumed the investment would be made in new pits, creating new jobs, and in the infrastructure of existing pits, to make them safer and guarantee their continued production. They had not bargained on the second focus of investment, in electronic control systems. Infrastructure investment — investment in new roadways underground, new face developments, and more up-to-date mechanical and electrical equipment — was essential in mines facing geological difficulties and years of underinvestment, as in South Wales, Scotland and the North East.

Even so, there still remains a contradiction within this investment. On the one side, the technology can help guarantee that a pit will stay open, producing coal. On the other side, the technology reduces substantially the number of jobs at the pit where it is applied; by hugely increasing productivity, it contributes towards the closure of collieries far away in other coalfields. It is this contradiction that the NUM finds so difficult to deal with today.

The clearest illustration of this problem can be seen in the Coal Board's intentions to use the remaining 25 million tonnes of high technology new capacity, agreed under 'Plan for Coal', to close 70 pits and cut 70,000 jobs, while maintaining the same deep-mined output of 100 million tonnes per year. Norman Siddall, caretaker Chairman of the Board for 1983, made this point at the industry's Consultative Committee in June 1983. As the new capacity was introduced over the next four to five years 25 million tonnes of older capacity would have to be phased out. And the new capacity will create many fewer jobs than it replaces; for example, at Selby less than 4,000 miners, deputies and management will produce nearly half of the new capacity output.

Microchip Mining

In September 1983, when Ian MacGregor was installed as Board Chairman, he inherited the most technologically advanced underground coal industry in the world. Not even the German industry can rival the diffusion of computer-controlled machinery and processes that can be found in half of Britain's 174 pits. And NCB automated equipment is being exported to the USA for installation in American mines, against the dominant flow of the electronics trade.

While on-line computer control systems have been and are being developed for every section of the coal industry, it is at colliery level that most effort has been made. Learning from a series of near-disastrous experiences with analogue electronics in the 1960s, Coal Board engineers were quick to adapt the newly available microelectronics to every major mining operation, both underground and surface.

The system that was developed became known as MINOS, an acronym rich in allegorical Greek myth, as indeed are several other system names. MINOS — Mine Operating System — is a standardized computer system for central control at collieries (Chandler). While it controls colliery operations centrally, using a closed loop design, its modularity means that different subsystems can be installed at different times, in a piecemeal manner. Later they can be linked together. Although the objective is to automate a mine as a working system, the piecemeal introduction of MINOS subsystems hid the process of its installation from miners' view. So, unlike the development of mechanized mining in the 1950s and 1960s, when miners played a major role in testing prototype machines and making innovations themselves, the new computerized systems were developed exclusively by control engineers and computer scientists, in many cases specially recruited from the aerospace industries, and based at the Mining Research and Development Establishment (MRDE).

MINOS subsystems have been developed to control or monitor in turn coal face machinery, underground transport systems, fixed plant like pumps and fans, the mine environment and coal preparation plants on the surface. With the exception of coal preparation plants these operations are monitored from a central control room on the surface. The first major battle over the new technology began over which union in the industry staffed these control rooms.

Initially the Coal Board placed members of the management union, BACM, and of the pit deputies' union, NACODS, in the new control rooms. The inter-union dispute was taken to the TUC Disputes Committee which decided in favour of the NUM. But it has become clear that, since the management unions lost the battle for control of the control rooms, management is now limiting the functions of NUM control room operators (Beynon, ed.). The closed loop systems design used by MRDE engineers has meant that many traditional mining skills are now incorporated into the computer systems, but the less well-defined 'management' skills associated with any overall skills are now being separated out and focused outside the central control room, in colliery engineers' or managers' offices, which are equipped with their own terminals.

The process of deskilling mining jobs, to a large degree assumed by the systems design adopted by the Board's engineers, is taking place throughout the subsystems. One area of colliery work that many miners originally thought would be immune from this process is electrical and mechanical maintenance. However, the IMPACT — Inbuilt Machine Performance and Condition Testing system, so far a separate system — is already starting to lead to the same polarization in skilled/unskilled. Small teams of specialized technical craftsmen, recruited from among mechanics and electricians, carry out the specialized technician level work associated with the new systems, while other craftsmen are finding their jobs being deskilled or are faced with the possibility that even under existing stringent mining safety regulations the NCB may be able to use unskilled workers to do their former work.

Of concern to the NUM, too, is a deterioration in the mental aspects of the working environment. While, as a rule, new mining technology does improve physical working conditions, the reduced autonomy of the deskilled jobs results in increased levels of stress. This has even been studied and is now recognized by NCB ergonomists.

However, by far and away the main threat to miners from new technology is the threat to their jobs. If the Board's intention is to reduce the industry to 100 pits, with 100,000 miners producing 100 million tonnes, over the next five years, it is conceivable that mining employment could be reduced further to 79,000, for only a slightly reduced output (Burns et al.). This is a figure also quoted by Tory MP Ian Lloyd, Chairman of the House of Commons Select Committee on Energy, a strong Board supporter. Curiously, it is one that the Coal Board reject as utter nonsense when it is made by NUM supporters or academics who have studied the issue independently.

The automation of deep mining in Britain, as one weapon to be used in battle with the miners, was not part of the original 1978 Ridley plan, but at management level inside the Coal Board it has been seen in that way for some time. While the drive to develop automated mining goes back to the Robens period, now it is much closer to fruition. At the same time the majority of people in Britain have been led to believe that coal mining — that first of all

major industries, and the power behind the industrial revolution
— is outdated, traditional and dying. Nothing could be further
from the truth. Few TV cameras have made the journey from the
deliberate dereliction of South Wales pits to the shopping complex
look-a-like superpits in the Selby countryside. But unless they do,
the fiction of a lame duck industry is maintained. This issue was
partly raised in a 'World in Action' programme, 'Miners and the
Microchip', when Professor John Ashworth was parachuted into
the Scottish mining community of Polkemmet. The NCB Scottish
Area wants to close Polkemmet Colliery, a move fiercely resisted
by the local NUM, even though they are in the middle of
Scotland's 'Silicon Glen'. In spite of the geography there are no
jobs in the burgeoning Scottish electronics industry for redundant
miners. For several years now it has become obvious that, apart
from university-trained systems staff, the only production jobs
that exist have been restricted to young female school leavers. By
the age of 21 even these young women are considered too old.

Ashworth has admitted this reality. As the National Economic
Development Council's Information Tehnology expert, he has
been pressing the Government hard over pursuing a more
vigorous development of the UK microtech industry. He has
suggested lamely that the miners get into university, train as
systems engineers and move to London! Perhaps even more
significant, Ashworth himself appeared to be totally ignorant of
developments taking place in the British coal industry and how
the automation of new and existing mines in the central coalfields
is being used to close pits like Polkemmet.

The NUM has been becoming increasingly aware of these
problems, but, like all other unions so affected, is finding it very
difficult to deal with the contradictions involved. From a
traditional position of demanding more investment to stop
colliery closures, thus preventing neglected pits slipping down the
low productivity spiral into an 'uneconomic' status, it is having to
face the fact that Selby will make all other pits appear uneconomic
by comparison. Projected net profits from the Selby complex,
when on full stream in five years, will be £17/tonne in 1983 prices.
The total annual *net* profit will therefore fall within the range of
£170–£212.5 million, based on an annual output of 10–12.5
million tonnes. In crude terms Selby's 3,500–4,000 miners are to be
used to replace 38,000 miners and the six-pit complex to close 45 of
the 'least profitable' existing mines.

On the other hand, using a strategy which uses *'free disposable
time'* as a measure of wealth, rather than labour time, Selby's

profits could be used to pay for a four-day week for 125,000 miners. Kept in employment would be at least 25,000 of those to be displaced by Selby. (This calculation is based on miners' average wage in 1983 of £164 per week.)

Accordingly, the central demand in the Draft Technology Agreement drawn up by the Union is for a four-day week, with no loss of pay. It would be practicable, when all other developments are taken into account. It also connects with the vision of a totally new form of society in which the employed are increasingly liberated from work and the unemployed are liberated from poverty and despair. Radical reductions in working time of this kind have been discussed in a theoretical way by socialists like Rudolf Bahro and André Gorz. In the context of the British mining industry, but with another eye fixed on the experience of the German Metalworkers' strike for a 35-hour week in 1984, we can see the practical shape such reductions might take.

In the meantime the forces arrayed against such a solution have steadily grown more determined and perhaps even stronger. However, with the strike over, the war itself remains to be won. Although 70 pits are threatened by automation and by the Government's refusal to allow British coal the market justified by its price and quality, they cannot be closed overnight. Within the next five years this battle over new technology in mining will continue, and an alternative marketing strategy is likely to be developed. These two issues can be more fully developed and linked strongly to the traditional struggle to defend mining communities.

References

H. Beynon, ed., *Digging Deeper: Issues in the Miners' Strike*, NLB/Verso, 1985.

A. Burns et al., 'Second Report on MINOS', Working Environment Research Group, Bradford University, 1984.

K. Chandler, 'Minos — a computer control system for collieries', 2nd International Conference on Centralised Control Systems, London, March 1978.

CSE Microelectronics Group, *Microelectronics: Capitalist Technology and the Working Class*, CSE Books, 1980, Chapter 10 on mines.

Ferrymoor Ridings NUM, 'Plan for Coal — With the NUM', 1985.

NUM, 'New Technology in Mining' (Briefing Booklet No. 6), available from St James House, Vicar Lane, Sheffield, S. Yorks.

STAR WARS/ EARTH WARS

Vincent Mosco

> Space for peaceful purposes – what a bunch of goddamned bullshit that was!
> – US Air Force General Bernard Schriever, 1983

Militarizing Communications Technology

The US military has shaped the electronics and communications industries in this country. The creation of the radio broadcasting system was a direct reponse to concern that the US could not challenge British imperialism, at least not in global communications, since Britain controlled the dominant means of communication – undersea cable. This control enabled British military and industry, from wool traders to news gatherers, to overwhelm competitors. To fight British domination, US government and leading businesses, including AT&T, General Electric, Westinghouse, and United Fruit, agreed to establish what the government called its 'chosen instrument' in communications, the Radio Corporation of America (RCA). RCA would be a cartel owned by the four corporate participants and an admiral would serve as government representative to the board. In this way, US military as well as industrial interests were maintained (Barnouw).

Even after the cartel broke up and RCA with its broadcast subsidiary, the National Broadcasting Company (NBC), moved on its own to establish dominance in the radio and television systems, the military continued to influence corporate policy. RCA is consistently within the top twenty companies in the amount of money it receives in Pentagon contracts. The Pentagon has exerted similar influence among RCA's major competitors, the Columbia Broadcasting System (CBS) and the American Broadcasting Company (ABC).

Forty years after the advent of broadcasting, the US military was faced with what it saw as another challenge to its global hegemony. This time the challenger was not British, but Soviet, and the technology was communications satellites. The response of the US government was much the same as in the RCA case. In 1962, the US established another 'chosen instrument', the Communications Satellite Corporation (Comsat), to advance the military and industrial applications of satellite technology in the US and, of particular importance, to organize an international body to promote and manage the development of communications satellites worldwide. Like RCA, Comsat was organized as a cartel, half of whose shares rested with the major US international telecommunications corporations: AT&T, RCA, International Telephone and Telegraph (ITT) and Western Union International (WUI). Again, the military maintained a substantial role in the operation and oversight of Comsat: one analyst dubbed it 'the old soldiers' home', a reference to the prominence of retired military officers working for the company. Comsat carried out its international mandate by creating Intelsat, an organization of over 100 nations, excluding the USSR and many of its allies. Comsat was able to construct the global system in such a way that policymaking power rested with nations based on the amount of their satellite usage. Consequently, one nation, the US, has consistently held over 20% of the voting strength in the organization. When we combine this with the votes of other advanced capitalist societies, the control of the US and Western Europe is ensured (Kinsley, Mosco).

In the last year, the computer industry received its first chosen instrument. Microelectronics and Computer Technology Corp. (MCC) is a company formed by the major data processing companies to coordinate international research and development activities. The ten companies listed as founding shareholders are Advanced MicroDevices, Control Data Corporation, Digital Equipment, Harris, Honeywell, Motorola, NCR, National Semiconductor, RCA and Sperry Univac. That the company represents a Justice Department-approved cartel of the top US computer firms is enough to raise a few eyebrows. In addition to this, the company is headed by Mr. Bobby Inman, former Deputy Director of the Central Intelligence Agency and the former director of the National Security Agency. The NSA is a top secret agency which operates a global computer/satellite system that routinely intercepts international telex, telegraph, telephone, radio and other transmissions emanating from or directed to the United States. It is anticipated that the MCC will provide the military, intelligence agencies, and the US computer/telecommunications industry with a vehicle to direct and coordinate international strategy (*Computerworld*, 31 January 1983).

The creation of MCC marks an organizational advance in the relationship between the military and the US computer industry. The relationship has always been a strong one. In the 1940s and 1950s the US government, led by the Pentagon, provided most of the funding for computer research. In addition the Pentagon provided big contracts to commercial firms to build the production equipment necessary to create the microchips that have revolutionized the industry. To complete the cycle, the Pentagon was the major consumer of computer products. Between 1958 and 1964 the military bought 35%–50% of integrated circuits produced in the United States (*Dollars and Sense,* October 1983, p.13). Today the Pentagon buys up about 20% of microchips produced by US computer firms (Mosco, *Pushbutton Fantasies,* p. 50).

In addition to shaping the commercial broadcasting, satellite, computer and related industries, the Pentagon controls a vast computer/communications network of its own. Indeed, the Pentagon and intelligence agencies control 25% of all radio frequencies used by public and private bodies in the United States. Other US government agencies control another 25% of frequencies and the remaining 50% are in the hands of commercial interests, chiefly large broadcasting companies. This is the rough equivalent of saying that one out of every four radio outlets in the United States is under the control of the US military. The Pentagon is thus the largest single user of telecommunications in the United States. Its annual budget for communications and intelligence is approximately $20 billion, or more than the total annual revenues of the over 8,000 commercial radio and 1,000 commercial television stations in the US. Through the Defense Communications Agency and the Defense Telecommunication and Command Control System, the Pentagon deploys a global system including its own and privately owned satellites, undersea cables, computers and other ground facilities to operate the most powerful communications and data processing system in the world (Mosco, *Pushbutton Fantasies*).

New Technology for the New High Ground

This system has enhanced the power of land, sea and air forces. Today, the Pentagon's computer/communications power is directed to opening up a new battlefield – outer space. In what *Business Week* magazine calls 'the most radical strategic policy change since World War II', the US is designing, building, and testing weapons that, until recently, most people believed existed only in movie theaters (*Business Week*, 20 June 1983, p. 52). But it is not simply the weapons themselves that reflect a policy change. More than this, it is the shift from using computer/satellite systems to support land, sea and air forces to making space itself the scene of battle. As one Air Force General explains it, 'Now what's so great about keeping our space inviolate from warfare? I can't see it... If we could get wars out in space we'd be a hell of a lot better here on earth' (*The Nation*, 9 April 1983, p. 444).

To build a space battlefield out of its lead in computer, communications and space technology, the US is bringing together leading experts and the most advanced facilities in the corporate and academic worlds. According to the President's chief science adviser, 'this is something that is going to be of the magnitude of Sputnik and Apollo and have even more impact' (*Business Week*, 20 June 1983, p. 50). Space war teams are integrating several fields of research – super computers, satellites, lasers, nuclear technology and others – to create new weapons systems. The stakes are especially high for leading military contractors because recent economic recessions have rolled back some of the growth they experienced in the 1960s. As a result, the Pentagon's industrial partners are providing public relations support for the military effort to win public backing and thereby make space 'the new high ground' (Graham).

Proposals to wage war in outer space attracted considerable media attention following President Reagan's 23 March 1983 address in which the President called for a national commitment to develop an orbiting fortress of space weapons to shoot down enemy missiles. Reagan's address was widely referred to as his Star Wars speech, though the label misses the important ideological tone that the President and his advisers have sought to establish. Far from praising the space-age weaponry that gave the film 'Star Wars' so much of its appeal, Reagan spoke of ending the terrifying threat of nuclear holocaust.

> What if free people could live secure in the knowledge that their security did not rest upon the threat of instant US retaliation to deter a Soviet attack; that we could intercept and destroy strategic ballistic missiles before they reached our own soil or that of our allies?
>
> Would it not be better to save lives than to avenge them? Are we not capable of demonstrating our peaceful intentions by applying all our abilities and our ingenuity to achieving a truly lasting stability? I think we are – indeed we must (*The New York Times*, 24 March 1983, p. 20).

Illusory as this promise is, and most analyses show that it is quite illusory, the Reagan proposals have a strong popular appeal.

This is because they offer the first opportunity, since the Soviet Union developed a nuclear capacity, for the US to end the danger of nuclear attack without having to trust the complexities of treaty arrangements. Consequently, even among people who might otherwise be sceptical, there is a desire to accept the nuclear shield concept because it promises to end the major fear of the last half of this century.

Details of the President's proposals are contained in recent military reports. The official Department of Defense directive for national strategy, the *Five-Year Defense Guidance*, outlines a plan to 'wage war effectively' from outer space (*Defense Week*, 28 June 1982, p.9). Two Air Force reports recently called on the US to establish military 'space superiority' (*Air Force 2000*, June 1983). To realize this vision of superiority, both the Air Force and the Navy have established Space Commands to co-ordinate their respective military space activities. In fact, the Air Force is calling for a unified Space Command that, in the words of Air Force Chief Verne Orr, 'Would treat space more properly as a fourth medium of defense' (*Defense 1983*, January, p.10). Research labs are being integrated in a billion-dollar Consolidated Space Operation Center at the home of the Air Force Space Command in Colorado.

In October 1983 a White House taskforce led by Defense Secretary Weinberger and William P. Clarke, then assistant to the President for national security matters, recommended immediate development of a space missile defense system. The White House body, the Defense Technologies Study Team, proposed spending $18–$27 billion over the next five years on a variety of technologies, including space laser systems, so that 'the potential for ballistic missile defense can be demonstrated by the early 1990s' (*Aviation Week and Space Technology*, 17 October 1983, p.16). The long-range goal is to have a multilayered ballistic missile defense in place within twenty years at a cost estimated at between $250 and $500 billion (Burrows).

Seizing the New High Ground

The idea of using outer space for military purposes is far from a new one. In fact, the US military interest in space predates the launch of Sputnik, the first Soviet space satellite, by over a decade. In 1946 the Air Force sponsored a Rand Corporation study, *Preliminary Design for an Experimental World-Circling Spaceship*. This report includes material on the military application of space satellites, including reconnaissance and communication (Rand

Corporation, 1946). Early Air Force projects on pilotless space-craft for communications, early warning, navigation and recon-naissance grew out of Rand's work. Research in the early 1950s stressed the need to develop an accurate and reliable Intercon-tinental Ballistic Missile (ICBM). This expanded in 1956 to fully-funded programs for satellite reconnaissance, and subsequently for climate assessment and, of particular importance, military communication.

The growth of US military forces worldwide created enormous problems of co-ordination and integration that only a very sophisticated communication system would be able to manage. Established systems for military command, control, communi-cations and intelligence (C^3I) were not sufficient, particularly in the event of a nuclear war. Space communications systems anchored in satellites would meet this pressing need. Con-sequently, the Pentagon began a major program to promote the development of communication satellites. Though these pro-grams were proposed by the Republican Eisenhower Adminis-tration, they received widespread backing from the Democrats. A 1957 Democratic policy statement signed by former President Truman, Senators Adlai Stevenson and Hubert Humphrey and New York Governor Averell Harriman reads:

> Let us not fail to understand that control of outer space would be a military fact of the highest importance . . . the air war of yesterday becomes the space war of tomorrow. We must do more than merely catch up. The all-out effort of the Soviets to establish themselves as masters of space around us must be met by all-out efforts of our own (Manno, p. 39).

And from Senator Lyndon Johnson, who would go on to become the President most deeply committed and closely identified with the space program:

> control of space means control of the world, far more certainly, far more totally than any control that has been achieved by weapons or by troops of occupation. Space is the ultimate position, the position of total control over Earth (Manno, p. 39).

In 1958 the Eisenhower Administration gave primary respon-sibility for manned space exploration to an ostensibly civilian agency, the National Aeronautics and Space Administration (NASA). Amendments to the NASA act subordinated the Administration to military requirements. Hence, the military was able to influence program decisions on the Mercury and Gemini flights that culminated in the Project Apollo lunar landings. In fact, the Air Force Systems Command won direct control over experiments on Project Gemini.

The Air Force co-ordinated its role in testing how to conduct manoeuvre and rendezvous operations in Project Gemini with its own program to develop military space vehicles. The major early projects were SAINT (for SAtellite INTercept), the first anti-satellite spacecraft, Dyna-Soar, a rocket-launched space glider developed for the Air Force by two former advisers to Hitler, Walter Dornberger and Kraft Ehricke, and BAMBI (for ballistic missile boost intercept), the earliest version of the current set of proposals to defend against ICBMs. BAMBI would deploy hundreds of satellites armed with heat-seeking missiles to be fired at Soviet ICBMs during their period of ascent. Mounting budget pressures, including increased military spending in Indochina, forced cancellation of these programs in 1963. Later on in the decade, a proposed Manned Orbiting Laboratory met a similar fate after $1.3 billion was spent on the project.

In the 1970s US military and intelligence agencies built satellite systems to enhance ground-based military forces, principally for reconnaissance, early warning, communication and navigation. In 1972 supporters of a strong military space program applauded President Nixon's approval of the Space Shuttle project. Though the Shuttle is formally under NASA control, its major characteristics, reusability, maneuverability, and large carrying capacity, are designed to precise military specifications. Moreover, unlike other NASA space projects, Shuttle flights are covered in secrecy and flight details are classified. According to the government's General Accounting Office, the military will require 49% of all flights through 1994 (US General Accounting Office, p.5). Most analysts agree that this is the price that NASA had to pay to maintain formal authority over the Shuttle and, indeed, to remain in existence (Karas).

A look at the pattern of spending for space activity reveals the extent to which the military has taken charge. Budget allocations support the contention of the director of the Congressional Office of Technology Assessment, who maintained in 1982 that with respect to outer space 'the military/intelligence community' has been given unambiguous control over policy decisions' (US, House, p.63).

Since 1958 the US has spent $150 billion on space programs. About $50 billion of this directly supported military space activity. In 1982 the formal allocation for Department of Defense space projects ($6.4 billion) surpassed the NASA budget ($5.9 billion) for the first time. The disparity grew in 1983 when DOD's space budget jumped 33% to $8.5 billion and that of NASA rose 15% to $6.8 billion. Moreover, the government's General Accounting Office estimates that 25% of NASA's allocation actually is in direct support of military space applications. For 1984, the Reagan Administration has increased the Pentagon space budget by about 9% to $9.3 billion and the NASA budget by 4% to $7.1 billion. If we add to the Pentagon total the 25% of NASA's budget that supports military programs and the estimated $3.4 billion that supports reconnaissance satellites and other classified programs, then total military spending on space for fiscal year 1984 amounts to $14.1 billion of all money spent on space activity (Center for Defense Information, 1983, p. 2). It is hard not to agree with Air Force General Bernard Schriever, who was put in charge of Air Force space programs in 1954 and whose ideas have propelled the Reagan Administration along its Star Wars track: 'Space for peaceful purposes – what a bunch of goddamned bullshit that was!' (Manno, p. 158).

Military Space Systems: Force Enhancement

The foundation of the military space program is the use of satellites as *force-enhancers*, space systems that are not weapons in themselves but enhance the capacity of land, sea, and air forces. The earliest military satellite applications include *reconnaissance*, *warning*, and *weather forecasting*. The US Air Force and the CIA opened the world's first reconnaissance satellite program in 1960. Since that time, both the US and the Soviet Union have launched hundreds of satellites that are placed in polar orbits which permit full coverage of the earth's surface every 12 hours. These data-gathering systems have been credited with providing verification necessary to make possible the 1963 Test Ban Treaty and Salt I. On the other hand, US reconnaissance satellite systems have been used to gather information on opposition movements within the United States. In a report on the highly secret National Reconnaissance Office, an agency whose budget is hidden in Air Force operations, *The New York Times* revealed that the NRO used satellites in the 1960s and 1970s 'to photograph antiwar demonstrations and urban riots, in an apparent effort to determine crowd size and the activity involved' (*The New York Times*, 1 March 1981, p.12). Furthermore, sophisticated reconnaissance systems buttress claims that it is possible to survive and perhaps even win a nuclear war. Satellite systems that make it possible to pinpoint and track nuclear and other weapons – on land, in the air, on and under the sea – fuel the hopes of those who would have the US strike first in order to gain the advantage in what is viewed as the inevitable nuclear confrontation (Aldridge).

Military *communications satellites* support these data-gathering systems. Two-thirds of US military traffic is routed through the Pentagon's own satellite systems. The military is now equipping these satellites with what the Pentagon views as the ability to resist jamming and interception *and*, most importantly, the ability to operate in a nuclear war. One new satellite system, Milstar, is under development largely to survive a nuclear war. Milstar will replace an effective Air Force satellite system that has provided effective communications links since 1974. Milstar uses small extra-high frequency terminals and antennas to extend the Afsatcom system to tactical nuclear weapons. According to the Defense Department, Milstar is 'to provide survivable and enduring command and control . . . through all levels of conflict, including general nuclear war' (*The Nation*, 9 April 1983, p.436).

Navigation and geodetic satellites enhance the capacity to sense an enemy and respond instantaneously. The heart of US activity here is the network of Global Positioning System (GPS) satellites. This system, also known as Navstar, uses an atomic clock to position forces all over the world on a common space/time grid. GPS applications appear endless, everything from helicopter blind landings to ultra-accurate bombing. One particularly destabilizing application, dubbed IONDS, calls for using GPS satellites to provide data instantaneously on the effectiveness of an initial nuclear strike in order to make a second wave attack far more productive. According to one analyst: 'With IONDS, a first strike intended to disarm the other side's ICBMs looks feasible with a far smaller arsenal than before' (Karas, 1983, p.138).

Military space applications for enhancing earth-bound forces have been around from the earliest days of the space program. Recent technical developments have added a new dimension to the significance of these systems. The ability to conduct universal reconnaissance, to survive a first-wave nuclear strike, and to position forces worldwide, have made it more realistic for military strategists to plan for prevailing in a nuclear exchange. Indeed, programs such as Milstar and IONDS have been created for just this purpose.

Anti-Satellite Weaponry

As powerful as force-enhancing systems are, they remain ancillary to ground, sea, and air forces. In addition to perfecting these satellite networks, the US now aims to 'project force in and from space' (*Defense Week*, 28 June 1982, p.9). The first step towards realizing this goal is to create an anti-satellite weapon (ASAT), for which the Reagan Administration has proposed spending $242.8 million in 1985. For the short term, the US is working on an ASAT that is essentially a one-foot-long cylinder. This Miniature Vehicle or MV ASAT is to be carried to an altitude of 60,000 feet by an F-15 aircraft. At that point a rocket would launch the MV to a target satellite. Infrared sensors locate the target and the MV rams it at high speed. The entire operation takes no more than ten minutes.

The Pentagon itself considers the MV ASAT far superior to its Soviet counterpart, which takes three hours to intercept its target and is limited to satellites under a 1000-mile orbit. In the June 1983 issue of *Scientific American*, three defense specialists write that the 'current Russian anti-satellite system presents a ponderous, inflexible, and quite limited threat to the United States'. They cite this as the 1979 opinion of the Air Force Chief of Staff and 'nothing has happened in the Russian test program since then to alter this 1979 assessment'. Hence, the Soviet ASAT cannot hit critical US communications satellites and early warning satellites which are stationed over 22,000 miles above the earth (Union of Concerned Scientists, 1983). According to the Director of the Pentagon's major research and development arm, the Defense Advanced Research Projects Agency, it will take the Soviet Union from five to ten years to counter the MV ASAT. Moreover the Air Force is working on new generations of ASATs that at minimum will increase their reliability against Soviet early warning and communication satellites stationed in highly elliptical orbits. Tests on later generation ASATs draw on the development of directed energy weapons such as laser beams, microwaves, and such nuclear-based weapons as subatomic particle beams which can destroy a target 20,000 miles away in one-tenth of a second (Jasani, p.435). Of these, laser weaponry offers the most near-term promise. The Pentagon plans to test components of its laser programs aboard the Space Shuttle in the 1980s. In October 1983 the Administration's Defense Technologies Study Team called for early development of several different laser systems. Such plans make it hard to disagree with those who suggest that the Space Shuttle itself is an ASAT weapon.

Ballistic Missile Defense

Interest in lasers and more advanced technologies is not limited to ASAT applications. In fact, the ASAT program is linked to a wider effort to develop a space-based ballistic missile defense system (BMD). In spite of the 1972 Anti-Ballistic Missile Treaty (ABM), which bans the development of space-based ABM systems, the US now spends $2 billion annually to research ABM technologies. In fact, in June 1984 the US for the first time successfully destroyed a test missile outside the earth's atmosphere with another missile. President Reagan's Study Team calls for increasing spending to about $26 billion over the next five years. Such amounts, large as they appear, dwarf in comparison with Pentagon estimates of what entire systems would cost. A multilayered ballistic missile defense system would take twenty years and cost between $250 and $500 billion to construct (Burrows, p.843).

The basic idea of a BMD system is to use directed energy weapons, such as lasers or particle beams, and high-speed pellet guns to attack Soviet ICBMs during every stage of their 30 minute flight. The first and most critical layer of defense is the attack against ICBMs shortly after they are launched. This boost-phase interception is significant because it destroys a missile before it has released its ten or so independently targeted warheads. But this means developing laser or particle beam weapons that can successfully reach about 2,000 boosters within five minutes of their flight. The second wave of defense would involve attacking the released warheads for the twenty minutes it takes them to reach their targets. The last stage, aptly named 'terminal defense', is the final effort to attack the warheads that make it through the first two layers by firing nuclear-tipped rockets directly at them.

In what it calls the Strategic Defense Initiative, the Reagan Administration plans to study both space and earth-based directed energy defense systems. Space systems would require the construction of a fleet of about 400 permanently orbiting battle stations, each weighing about 100 tons. Systems based on earth would nevertheless require a set of orbiting mirrors or mirrors launched from ICBMs (called pop-up mirrors) that would warn of an attack and at which the earth-based energy beams would fire. The mirrors redirect the energy beams at the rising enemy boosters and released warheads (US Office of Technology Assessment).

As amazing as descriptions of Ballistic Missile Defense systems appear, there is no lack of confidence among the scientists working in the BMD. A *New York Times* reporter interviewed several in a visit to the center of BMD research, the Lawrence Livermore National Laboratory. Though quite young (the oldest of the participants was 34), they play major roles in the development of Star Wars weaponry. One typified the views of the

Livermore group:

> I don't think I fall in that category, of working on weapons of death. We're working on weapons of life, ones that will save people from the weapons of death.
>
> There's almost an infinite number of issues to be pursued. The number of new weapons designs is limited only by one's creativity. Most of them have not been developed beyond the stage of thinking one afternoon, "Gee, I suppose you can do so and so." There's a tremendous number of ways one might defend the country (*The New York Times*, 31 January 1984).

Corporate Victors

The enthusiasm of these young weapon researchers notwithstanding, questions abound in this area. But no one questions that a number of corporations benefit quite extensively from space military contracts. The electronics industry as a whole has grown in response to the growth in military spending initiated by the Reagan Administration. The Electronics Industry Association estimates that electronics now comprise 33% of the cost of aircraft, 45% of missiles, 66% of space systems, 22% of ships, 24% of vehicles and 88% of communications equipment. This makes military electronics a $44 billion industry that is anticipated to grow at 15% annually through this decade. Military communications will grow at the even faster rate of about 20% a year (Scholl, p.8).

The clear victors are led by Rockwell International, McDonnell Douglas, TRW, Ford Aerospace and Communication, Martin Marietta Aerospace and Boeing. Rockwell is the leading corporate beneficiary of the Space Shuttle and in the vanguard of a public relations effort to sell the idea of space warfare to a sceptical public. In 1981 alone, NASA awarded Rockwell $1.47 billion in Shuttle contracts. In that same year the Pentagon gave Rockwell another $1.12 billion in general defense contracts. The company responds by taking a decidedly military view of outer space. According to a manager with Rockwell's Space Transportation System Division:

> When people ask how much is enough for military spending they compare it to welfare or education or something else we're doing. I can't understand that...the United States could win in a technological competition with the soviets if we were committed . . . if we set our minds to it we could get a decisive advantage (Karas, p.59).

Rockwell promotes this view as a leader of the powerful lobby, the Aerospace Industries Association, two-thirds of whose members have Space Shuttle contracts. Rockwell's glossy promotional brochure, 'Space: America's Frontier for Growth, Leadership and Freedom', offers a century of space weapons development. All of this is capped in the 21st century by a fortress positioned 22,500 miles above the equator to conduct fool-proof worldwide surveillance and manage battles effectively under the protection of powerful laser systems.

McDonnell Douglas is responsible for what *The New York Times* refers to as 'some of the most golden military aircraft in the Pentagon's arsenal' (18 March 1984). *The Times* refers to the F-15 for which the Reagan Administration has requested $2.2 billion for 48 planes in 1985. The aircraft, the 'launch' vehicle for the US ASAT weapon, has earned McDonnell Douglas about $1.5 billion annually in the 1980s. The company has already benefited from ballistic missile defense contracts, earning $180 million in 1982.

Will It Work?

Major questions have been raised about whether this leap into space warfare will accomplish what its proponents claim - protection against a nuclear attack from the Soviet Union. The development of an ASAT system will lead the Soviet Union to push ahead with its own anti-satellite weaponry. In fact, a number of analysts have pointed out that the US has the most to lose from an ASAT build-up because more of its defense depends on satellite systems (Union of Concerned Scientists, 1983).

Similar questions have been raised about ballistic-missile defense systems. Even the $½ trillion cost estimates for a complete system do not take into account the need to respond to inevitable countermeasures. The congressional Office of Technology Assessment (1984) has considered a range of powerful responses, including the use of anti-satellite weaponry to attack the mirrors and sensors required of a BMD system, fast-burn boosters to limit the effectiveness of boost-phase interception, multiple decoys and shielding. According to Robert Bowman, former Air Force Director of Advanced Space Programs Development and Manager of Advanced Space Programs for General Dynamics, lasers are particularly susceptible to a variety of countermeasures. These include spinning the ICBM to distribute the laser energy, using burn-resistant surfaces, deploying decoys, or coating the missiles with a mirror-like surface to reflect laser energy (Bowman, p. 5). Karas (1983) reports on how the Voigt Corporation, in research for the US Navy, has developed a highly-polished aluminium surface that reflects 97% of the energy from infrared laser beams. As a result of these findings, the Office of Technology Assessment

(1984) concluded that 'for every defense concept proposed or imagined, including all of the so-called Star Wars concepts, a countermeasure has already been identified'. Considering the nullifying power of countermeasures, even proponents admit that there can be no guarantee that a ballistic missile defense system will destroy all incoming missiles (Carter and Schwartz). As a result, the Office of Technology Assessment concludes:

> The prospect that emerging Star Wars technologies, when further developed, will provide a perfect or near-perfect defense system, literally removing from the hands of the Soviet Union the ability to do socially mortal damage to the United States with nuclear weapons, is so remote that it should not serve as the basis of public expectation or national policy about ballistic missile defense (BMD). This judgment appears to be the consensus among informed members of the defense technical community. (p.81).

The Financial Cost of Star Wars

Those who question whether the Star Wars defense will actually succeed tend also to criticize the tremendous cost of these weapons systems. According to a former Air Force and General Dynamics systems engineer:

> Responsible Pentagon analysts estimate the true cost at anywhere from $100 billion to over a trillion! It should also be remembered that the 432 huge battle stations in the "initial system" represent just the nose of the camel. The second-layer system would require a follow-on to the shuttle, a manned space station and repair facilities. Kosta Tsipis of MIT estimated that it would take $100 billion just to launch the fuel into orbit for the second-layer laser system (Bowman, p. 6).

According to a report prepared for the Union of Concerned Scientists, chemical laser 'battle stations' in low orbits or the equivalent of 'space trucks' carrying weapons systems will have to number in the thousands in order to provide adequate coverage of Soviet silos. Simply to loft these stations into orbit would cost over $70 billion. Moreover, excimer lasers on the ground, whose beams would be reflected by a thousand or so orbiting mirrors, would require power plants that alone would cost upwards of $40 billion (Union of Concerned Scientists, 1984, pp.2–3).

Supporters of the Strategic Defense Initiative are more graphic in their recognition of the enormous costs involved. According to the Under Secretary of Defense for Research and Engineering, just the research and development portion of the ballistic missile defense program has at least eight components, 'every single one ... equivalent to or greater than the Manhattan Project' (Union of Concerned Scientists, 1984, p.3).

Large as these estimates appear, none take into account the obvious countermeasures that the Soviet Union would have to take in order to defend itself against the possibility that a combination anti-satellite weapon system and ballistic-missile defense would increase the likelihood of a US first strike. It is apparent that top US officials recognize that such a fear is understandable. Defense Secretary Weinberger has said that he would view a similar Soviet system as 'one of the most frightening prospects' imaginable.

Critics of the militarization of space point to research that indicates the negative effect of defense spending on employment and the general level of spending on such basic needs as housing, transportation and essential social services. They argue that a commitment to a new military system will erode the US competition position in the world economy and make it increasingly difficult to maintain the support of people whose schools, health centers, welfare offices, etc, are cutting back services or closing down entirely because the government chooses to build orbiting laser battle stations (Council on Economic Priorities).

Launch on Warning

Star Wars is not the only film reflected in the plans of military strategists. In the summer of 1983 the movie *War Games* raised the spectre of a computer-triggered nuclear war. While it is unlikely that a teenage video game enthusiast would be, as the film portrays, the nuclear culprit, the film does accurately depict the danger of a 'launch on warning' strategy. Launch on warning is a policy of using nuclear retaliation against an enemy attack indicated solely by computer-satellite-radar tracking systems. The policy of relying on machines to launch a nuclear war is, in the minds of some, the logical response to a decision-making time frame now measured in minutes and seconds. Specifically, the nuclear distance between the US and the Soviet Union is about thirty minutes, between Europe and the USSR less than ten.

There is growing concern that launch on warning widens the potential that a mechanical error will lead to nuclear war. This concern has been expressed by some at the command posts of the US military industrial complex. In an interview with *Computerworld* magazine, Thomas J. Watson, Jr., the head of IBM, initially denied that there was much likelihood of war begun by a mechanical malfunction, but appeared to change his mind in the course of his answer:

. . . the more the whole philosophy of launch-on-warning becomes attractive, the greater the danger. And as machines of war and missiles become more prey to pre-emptive strike, the more temptation there is to put more and more of the data in the hands of the computer and take the human being out of the equation. To the extent that you do that, you are indeed putting the US into a position where a computer could trip us up pretty badly (*Computerworld,* 15 June 1983, p.15).

General Richard Ellis, who was once reponsible for the US major nuclear force, including bombers, tankers, reconnaissance aircraft and ICBMs, expressed a similar concern about reliance on machines in situations of enormous time pressure. At a Harvard seminar, he describes what he calls the 'iffy business' of relying on processed information to make judgements about a possible attack:

> All the information comes into NORAD (North American Air Defense Command). It's ground up in their computer programs and presented to them in a matter of minutes, in some cases seconds, as fused information, which indicates to the commander out there that such-and-such is happening. All one can do is hope that the software isn't faulty, or the hardware isn't spooky, and the person is not making a hasty judgement. Things can go wrong (Harvard University, p.5).

In addition to their concern about mechanical malfunction, Pentagon officials are concerned that it is not possible to prevent violations of computer security. Despite a $100 million effort to make impregnable computers that process sensitive information, violations were reported with regularity last summer when a group of young computer hackers, as they are called, successfully keyed into a computer that processes non-secret information at Los Alamos National Laboratory, where nuclear weapons are designed (*The New York Times,* 25 September 1983).

This concern about mechanical malfunctions takes an overly simplistic view of responsibility. There is a series of very direct human decisions that enter the chain leading up to a decision to launch the missiles. Simply because the last step in the decision process went awry because of a mechanical error is not sufficient reason to assign responsibility for a nuclear war to a set of machines. Nowhere was this temptation acceded to more, and nowhere more irresponsibly, than in the Vietnam War. In that conflict, computers in the field were specifically programmed to tell Pentagon computers that raids over neutral Cambodia were actually raids over Vietnam. Highly-placed elected officials who were permitted to see the summaries produced by the Pentagon computers wrongly believed they were receiving a privileged insight into field action. Of particular interest is the response of then chairman of the joint Chiefs of Staff, Admiral Moorer, to a congressional committee investigating the matter. According to the Admiral, 'It is unfortunate that we had to become slaves to those damned computers' (Weizenbaum, p.560). Launch on warning and computer malfunction are far more complex matters when we consider that sitting near the very top of the nuclear chain of command is someone who programs computers to mask a war and then claims to be enslaved by these same computers.

Nevertheless, it is clear that the new wave of Star Wars weaponry makes decision-making more complex and more than ever under the pressure of split-second time spans. This is the case with both anti-satellite and ballistic missile defense systems. As Bowman points out, if a major warning satellite belonging to one side were to cease working because of a meteor strike or electrical failure, the side with the malfunction or destroyed satellite might conclude that an ASAT had done it and that this is the prelude to an attack. Does it spend precious time checking out the malfunction or does it launch a 'retaliatory' strike? ASATs would obviously aggravate the uncertainty that already exists.

Similarly, ballistic missile defense systems require a leap in reliance on machine judgement. As Steinbruner points out, it now requires about two minutes just to process and verify the data from early warning satellites; it takes an additional 200 seconds to loft an interceptor from a submarine to the firing point. This is much longer than the 155 seconds during which the MX missile boosts. Perhaps the complex sequence of computations and decisions can be arrived at in a matter of seconds provided that no humans have taken part in the process (*Scientific American*, January 1984 and Union of Concerned Scientists, 1984, p.45). In other words, greater reliance on BMD systems means greater reliance on non-human decision making.

First Strike

A number of critics argue that the major worry is not an accidental firing of offensive or defensive weaponry, whether triggered by human or machine. The real concern is that the development of ASAT and BMD systems are preludes to a calculated first strike aimed at eliminating the Soviet Union as a viable society. Critics who adopt this perspective argue that the scientists who have played a key role in persuading the Reagan Administration to pursue anti-satellite and ballistic missile defense systems are no fools. They realize that no ASAT or laser BMD system will prevent all Soviet missiles from reaching the United States. They further realize that even an optimistic failure rate of 2–5% would be catastrophic for the United States. Then, in spite of all the questions raised, why pursue the Strategic Defense Initiative? First Strike theorists like Michio Kaku, professor of nuclear physics at the Graduate Center of the City University of New York, claim that 'the laser ABM (Anti-Ballistic Missile), with all of its limitations, may have effective applications in conjunction with the launching of a pre-emptive first strike' (*The Progressive*, June 1983, p. 22).

A US first strike would be coordinated by the force-enhancing systems described earlier, hardened to survive a protracted nuclear war. A first strike would be led by ASATs to destroy the Soviet satellite early warning and communications capacity. Land-based MX missiles and the submarine-launched Trident II will drop hydrogen bombs on each of the Soviet SS-18 and SS-19 missile silos. A BMD system will be geared to the thousand or so warheads, principally from the Soviet submarine fleet, that manage to survive. The few missiles that might elude both a first strike and the ballistic missile defense provide the basis for the Reagan Administration's proposals for civil defense and massive relocation plans. According to Colin Gray, a State Department consultant and supporter of planning for a first-strike victory:

> The United States should plan to defeat the Soviet Union and to do so at a cost that would not prohibit US recovery. Washington should identify war aims that in the last resort would contemplate the destruction of Soviet political authority and the emergence of a postwar world order compatible with Western values . . . A combination of counterforce offensive targeting, civil defense, and ballistic missile and air defense should hold US casualties to approximately 20 million, which should render US strategic threats more credible (*The Progressive*, June 1983, p.22).

The goal then is not merely a first strike but a mix of first-strike attack, anti-satellite and anti-missile weaponry, and civil defense planning to guarantee that the US prevails over the Soviet Union (Aldridge).

The End of Treaty Commitments

Some would argue that the Strategic Defense Initiative need not necessarily be part of a first-strike plan in order to work in the Administration's interest. In fact, Star Wars systems themselves would not necessarily have to be effective weapons to satisfy a key interest of the President and his followers. For years a major segment of the Right in the United States has decried the agreements that the US and the Soviet Union have worked out to limit the potential for nuclear war. The Right has attacked the Test Ban Treaty, which forbids nuclear detonations in the atmosphere and in outer space, the Outer Space Treaty, which bans space-based nuclear weapons and weapons of mass destruction, and the ABM Treaty, which forbids the development, testing and deployment of space-based ABM systems and protects satellites serving as 'national means of verification' (Union of Concerned Scientists, 1983, p.2).

The Strategic Defense Initiative provides the Reagan Administration with an opportunity to sever these treaty obligations without any official pronouncements that might draw the ire of world opinion. Rather, the Administration is simply responding in a defensive way to the massive threat it perceives from the Soviet Union. Certainly, the congressional Office of Technology Assessment recognizes the consequences of a space buildup for US treaty obligations. In its report on BMD it lists 'Demise of the ABM Treaty' as a side-effect of BMD deployment:

> An arms control treaty obviously cannot serve as its own justification, and presumably virtually everyone would agree to the abandonment of the ABM Treaty the moment it ceased genuinely to serve the national security . . . As a practical matter, it is impossible to overturn the Treaty's technical provisions without calling into question US commitment to the whole fabric of the SALT/START process (p. 77).

The Reagan Administration has had a harder time than it may have expected in convincing people that it is gutting the treaties out of defensive necessity. This is owing in part to the consistent Soviet policy of calling for an end to the development of space weaponry. In August 1981 the Soviet Union submitted to the United Nations a draft treaty calling for a 'prohibition on the stationing of weapons of any kind in outer space'. The Soviets, under Andropov and Chernenko, have followed this up with persistent calls for opening negotiations for a general ban on space weaponry. The US has not responded at all to the Soviet draft treaty. In April 1984, the President presented a report to Congress on why his Administration refused to negotiate on this issue. The report was the price Congress exacted for freeing $19.4 million in funds for ASAT development. The report says that the Administration has not found any proposals for limiting space weapons that it considers 'in the overall interest of the US and its allies' (*Wall Street Journal*, 3 April 1984). Election year pressures have forced Reagan to soften his position a bit. The Administration now claims that it will present a set of proposals, none of which call for a total ban on space weaponry, within a short time.

According to a person to whom the *Washington Post* refers as a 'high-ranking official who opposes any negotiations on ASAT', 'What's happening is it's election year and the Administration is losing its nerve' (16 June 1984).

The Strategic Defense Initiative is consistent with an Administration policy of opposing multilateral agreements generally and in particular any sort of agreement involving the Soviet Union. The militarization of outer space offers it the opportunity to do so and at the same time lay the technical basis for a general self-reliance policy worldwide. The Reagan Administration appears to be saying that it is time for the US to give up the pretense of being able to guarantee imperialist interests throughout the world and instead must define its interests solely in national terms and in competition with other imperialist nations.

Some observers are recognizing that this position is not without its supporters in the Democratic Party. For example, Presidential candidate Gary Hart has called for a substantial cut in American troops in Europe and a complete revamping of the North Atlantic Treaty Organization. Moreover, he has said that European

The Great Communicator

countries should make their own military plans to secure their supplies of Persian Gulf oil. The difference between Hart and the Reaganites is that, where Reagan sees a militarized outer space providing the protective umbrella, Hart points to a massive buildup of US naval forces to protect what he refers to as 'America's sea lanes'.

Star Wars vs. Trilateral Security

As we have seen, there is no lack of critics saying that the Strategic Defense Initiative will not work, that it costs too much, promotes a first-strike strategy, increases the dependency on computer decision-making, and violates several effective treaties. But none of these arguments may be as effective in opposing Star Wars as the fact that a substantial segment of the US power structure opposes the militarization of outer space. More generally, this segment opposes the Reagan (and Hart) view that the US should develop an independent position worldwide, even if that means in strident competition with capitalist allies. In contrast to the Reagan view, the opposition, sometimes referred to as Trilateralists, hold that the US should continue to maintain its post-World War II role as promoter of a general capitalist consensus and protector of mutual interests.

The Trilateralist position is stated most explicitly in a report issued at the same time as the White House received a report from its Study Team calling for a commitment to the Strategic Defense Initiative. This report, sponsored by the elite Trilateral Commission, opposed the deployment of a BMD system. The report, *Trilateral Security*, argues that:

> Such a deployment could not only leave Western Europe and Japan exposed to Soviet Missile atacks . . . deployments beyond those permitted by the ABM treaty of 1972 would be viewed by many, both in Western Europe and Japan, as well as in the United States, as an abandonment of arms control's most effective achievement to date and as prejudicial to prospects for improved East-West relations (Smith, 1983, p.47).

The Wall Street Journal (4 April 1984) reports that there is growing concern among NATO officials that the US will abandon its commitment to European security in favor of investing billions in a system that is built to protect the United States. Both the Trilateral Commission and supporters of a strong NATO would like to see the US direct its defense dollars to significantly strengthened conventional armed forces.

The Reagan Administration has felt the pressure of this criticism. In hearings before the US Senate in April 1984, the director of the ballistic missile defense program testified that the Reagan Administration has decided to expand research to include protecting European allies from short-range Soviet missiles as well as protecting the United States against long-range attacks (*The Washington Post,* 25 April 1984).

In June the opposition stepped up the attack by forming the National Campaign to Save the ABM Treaty. One of the leaders of the group is Gerard Smith, chief US negotiator at the first strategic arms limitation talks, which produced the ABM treaty, and principal author of *Trilateral Security*. In addition to Smith, the National Campaign includes former President Jimmy Carter, former secretaries of state Dean Rusk, Cyrus Vance and Edmund Muskie, former defense secretary Robert McNamara, and former CIA directors William Colby and Stansfield Turner. Moreover, former Trilateral Commission member and Democratic Party Presidential candidate Walter Mondale supports this position. While it is difficult to foresee the outcome of this conflict, two points are clear. The debate over Star Wars is not merely about whether the US is to develop new military technologies. Rather, it is intrinsically tied to conflicting perspectives about the US role in the world. In that sense both Reaganites and Trilateralists employ *technicist* arguments. That is, they each mask political visions, different means of maintaining US imperialism, with technological claims. Moreover, the Trilateralists show that one can oppose Reagan's Star War visions and remain a strong proponent of US imperialism. Indeed, the *Business Week* reference to the Strategic Defense Initiative as a 'radical policy change' makes more sense when we see Star Wars as a departure from the US post-World War II position as leader and protector of advanced capitalist societies.

Opposing Star Wars

There is room to oppose the militarization of outer space without joining with the Trilateralists in calling for alternative forms of imperialist aggression. It is encouraging to note that a number of associations of scientists in the United States have organized to oppose, not just space war, but the militarization of science in general. Particularly notable among these is the Union of Concerned Scientists, whose reports on anti-satellite and BMD weaponry have provided the basis for organizing wider opposition constituencies. In addition, Science for the People, which publishes a magazine by that name, has consistently linked opposition to Star Wars to its general anti-imperialist position. Opposition has been slower to develop in the computer community.

Among the US groups that have taken a strong position against the militarization of their work are Computer Professionals for Social Responsibility and High Technology Professionals for Peace. Finally, among media scholars and activists, the Media Network and the Union for Democratic Communications clearly see the Reagan strategy as a dimension of American imperialism.

The Reagan Administration's plans to militarize outer space may succeed in arousing considerably more opposition. The significant question is whether that opposition is directed at the technology alone or, more widely, against the imperialist policy for which Star Wars technology is merely the instrument.

Bibliography

This article expands significantly on a piece that appeared in the March 1984 issue of *Le Monde Diplomatique*.

Robert C. Aldridge, *First Strike,* Boston, South End, 1983.

Erik Barnouw, *A Tower in Babel,* New York, Oxford, 1966.

R.M. Bowman.'The Fruits of Space Exploration: Cornucopia or Armageddon?', paper presented at the 34th Congress of the International Astronautical Federation, Budapest, Hungary. 10-15 October 1983.

William E. Burrows, 'Ballistic Missile Defense: The Illusion of Security', *Foreign Affairs* (Spring 1984), 843–56.

Ashton B. Carter and David N. Schwarz, eds., *Ballistic Missile Defense,* Washington, D.C., The Brookings Institution, 1984.

Center for Defense Information, 'Militarizing the Last Frontier: the Space Weapons Race', *The Defense Monitor* (August 1983).

The Council on Economic Priorities, *The Costs and Consequences of Reagan's Military Buildup,* New York, 1982.

D.O. Graham, *High Frontier: A New National Strategy,* Washington, D.C., 1982.

Harvard University, Program on Information Resources Policy, *Seminar on Command, Control, Communications and Intelligence,* Cambridge, Massachusetts, 1982.

B. Hasani, 'Space: Battlefield of the Future', *Futures* (October 1982), 435-47.

Thomas Karas, *The New High Ground,* New York, Simon and Schuster, 1983.

Michael Kinsley, *Outer Space and Inner Sanctums,* New York. Wiley Interscience. 1976.

Jack Manno, *Arming the Heavens,* New York, Dodd, Mead, 1983.

Vincent Mosco, 'Comsat Presidential Directors – Promise and Performance'. *Telecommunications Policy* (December 1981), 251–64.

Vincent Mosco, *Pushbutton Fantasies: Critical Perspectives on Videotex and Information Technology,* Norwood, N.J., Ablex, 1982.

Rand Corporation, *Preliminary Design for a World-Circling Spaceship,* Santa Monica, California, 1946.

Jayce Scholl, 'The Force Is With Them: Star Wars defense will benefit a slew of companies', *Barron's,* 30 April 1984.

G.C. Smith,P. Vittorelli, and K. Sacki, *Trilateral Security,* New York, The Trilateral Commission, 1983.

Union of Concerned Scientists, *Anti-Satellite Weapons: Arms Control or Arms Race?,* Cambridge, Massachusetts, 1983.

Union of Concerned Scientists, *Space-Based Missile Defense,* Cambridge, Massachusetts, 1984.

US Congress,House Committee on Science and Technology, Sub-committee on Space Science and Applications, 4 August 1982.

US Congress, Office of Technology Assessment, *Directed Energy Missile Defense in Space,* Washington, D.C., 1984.

Joseph Weizenbaum,'Once More, The Computer Revolution', in Forester, *The Microelectronics Revolution,* Cambridge, MIT Press, 1981.

MARXISM FROM ABOVE: Philosophy, Science and Professor Kolakowski

Jonathan Rée

Essay review of Leszek Kolakowski, *Main Currents of Marxism: Its Origins, Growth and Dissolution*, Three Volumes, Oxford University Press, 1978, 1981, Pb £3.95 each.

I keep making resolutions to read comprehensively in the classics of Marxism, starting with the *Collected Works* of Marx and Engels, then Lenin, Luxemburg, Trotsky, Lukàcs, Gramsci, Korsch . . . But there are so many Marxist classics on my bookshelves, staring back at me in reproachfully good condition, that I have to admit that these resolutions may never be fulfilled. *Main Currents of Marxism*, however, looks as though it may provide a short cut — three large volumes, first published in 1978, now available in paperback, festooned with such testimonials as 'a masterpiece', 'magisterial', 'the most complete and satisfying survey of Marxist thought ever written'. It is filled with critical digests of the great classics in the Marxist canon, and many minor ones too: surely, exactly what people like me have been looking for.

Its author is uniquely qualified for the task. His experience of writing a vast and authoritative book on reformation Christianity must give him a special skill in describing theories associated with turbulent social movements, and he is marvellously well-read in the main European languages. Above all, his own life has been inseparably and tragically tangled with the development of post-war Marxism. He was born in Radom in Poland in 1927, and joined the Polish Communist Party in 1946. He became a teacher of philosophy in the University at Lodz in 1947, and moved to Warsaw in 1950. He played an important part in the relatively exuberant intellectual culture of Poland in the mid-1950s and in

1957 began to publish a series of essays — notably 'Responsibility and History' and 'Permanent vs. Transitory Aspects of Marxism' (both translated in *Marxism and Beyond*) — whose celebration of intellectual and ethical independence offended Party authorities and led to him being denounced as a 'revisionist' by Gomulka. However, for the 'New Left' in America and Western Europe, his name became the symbol of a hoped-for alliance between dissident socialists on both sides of the 'Iron Curtain', an alliance to be founded upon the doctrines of 'Socialist Humanism'. In 1966, following a speech he gave to mark the tenth anniversary of the 'Polish October', he was expelled from the Party; and after student demonstrations in March 1968, he was deprived of his University post. He escaped to America but was appalled by what, to him, was the deluded and incipiently authoritarian pseudo-Marxist sloganizing of the student movement; nor was he reconciled to it when his car was burnt in California in reprisal against his alleged political cowardice. Since 1970 he has been a Fellow of All Souls College, Oxford: a tall, gaunt and wistful figure, walking meditatively with a splendid translucent cane, and gazing abstractedly into the distance.

Main Currents of Marxism includes only a couple of references to its author's own experience. On one occasion, he tells us that Polish revisionists 'made a devastating comparison between socialist reality and the values and promises to be found in the "classics" ': but, waiving the privileges of an autobiographer, he concludes that 'the proportion as between faith and deliberate camouflage is hard to estimate at this distance of time' (III, 457–8). And he employs the same austere and downbeat third person when, referring to his own involvement in Polish Marxism in the early 1950s, he says that 'he does not regard the fact as a source of pride' (III, 173). Such self-effacement is extraordinary when an author is dealing with events which have altered the course of his life; and it creates, for the reader, an enigmatic and touching authorial personality. This accounts, I suspect, for the extraordinary — and, as I shall argue, wholly misplaced — enthusiasm with which the book has been received.

ONE: A LIFE OF MARXISM

Main Currents attempts to frame the history of Marxism as a coherent narrative, in a form rather like that of an edifying Victorian 'life'. The ancestry of our protagonist is established in the first hundred pages, and then he is seen setting forth with unclouded youthful cheer in Volume One (subtitled 'The Founders'), carrying all before him in Volume Two ('The Golden Age'), and meeting his sorry fate in Volume Three ('The Breakdown'). As with any biography, the argument depends on the way the story is told, and especially on the adumbrations by which the narrator lets on that he is in the know and beckons the reader to join him there.

Marxism, we learn, began as an attractive blend of 'Romanticism', 'Prometheanism' and 'Rationalism', its tragic flaw; also its fatal charm, being its buoyant certainty that reality must confirm with its theory-begotten schemes, rather than the other, more sober, way round. The story begins seventeen centuries ago, with 'the soteriology of Plotinus' — a theory that human nature will find fulfilment only by fleeing temporal contingency and merging with the Absolute. Having thus located what he calls 'the origins of dialectic' in third-century Christianity, and whilst assuring us that this is not 'a piece of Hegelian exposition' (I, 17), Kolakowski tells of how 'the dialectic' was transmitted into the 'Christian theogony' of John Scrotus Eriugena (*sic*) in the ninth century: an idea, that is, of human history getting absorbed by the higher reality of a self-creating Absolute Being which, Kolakowski reveals, is none other than 'the pattern of Hegel's *Aufhebung* or "sublation" ' (I, 28–9).

With these Hegelian launch-pads installed at the beginning, the story rises irresistibly through the clouds of Romanticism and Hegelianism till we meet a cockeyed theologian, dottily nominating a fantasized proletariat as the agent of human reconciliation with the One. This is Marx, who in contrast with his companion Engels (an unphilosophical positivist, it seems) is consistently portrayed as a visionary millenarian, who saw his own time as the ultimate crisis of humanity's alienation from itself — an idealistic socialist for whom 'not poverty but loss of human subjectivity is the essential feature of capitalist production' (I, 287), and a nostalgic Golden-Ager who looked to Socialism for a 'return to a situation in which only individual human subjects truly exist' (I, 311).

Volume Two presents a race of lesser heroes, Marx's followers within the Second International, courageously defending a dangerously decomposing intellectual inheritance. Despite their differences, they all end up impaled on the same philosophical paradoxes of liberty and authoritarianism, because of their common attempts, vestigially but unconsciously theological, to set no store by the here-and-now, and to lay up their standards of conduct in some never-yet land of fully-realized subjectivity. In fact it was pre-eminently through the intellectual self-mutilation with which they sacrificed ordinary respect for brute facts on the altar of fantastic philosophical schematisms that they demonstrated their loyalty to Marx. Luxemburg, for example, was loyal insofar as 'her account of social reality was often based more on theory than on observation'; her views were 'deductions from a Marxist schema, with the minimum of correction in the light of experience' (II, 82, 93). But Jaurès, though a 'soteriologist', was not really a Marxist, since he 'never treated Marxism as a self-sufficient and all-embracing system, from which the interpretation of all social phenomena could be deduced . . . a metaphysical key to the universe . . . an all-or-nothing position . . . a key to all human problems' (II, 115, 120). Once in its stride, Kolakowski's narrative acquires a tiresome repetitiveness, each episode revealing yet another dreamer beguiled by Marx the miraculous locksmith. Thus Lafargue 'believed that as Marx had provided a universal key it could be used to unlock the secrets of all sciences however little particular knowledge he might possess' (II, 142). No true Marxist could conceive that 'history might have thrown up social forms which were not necessarily reducible to a single pattern', (II, 236) so the 'Austro-Marxists' were not really Marxists at all, since they 'did not regard Marxism as a closed, self-contained system' (II, 240). Plekhanov made up for this by establishing a style of thought 'which recalls the polemics of religious sects' (II, 344) and hence set the stage for Leninism: arbitrary Party power, quite at odds with Marx's individualistic vision of socialism, but, we are assured, a valid deduction from his 'historiosophical' longings for a secular apocalypse nevertheless.

Volume Three is even more depressing, portraying Lenin's successors tearing themselves apart in an orgy of intellectual vandalism; for instance, 'Lenin at least tried to argue, though his arguments were logically useless, but Bukharin has not even this to his credit' (III, 61). Predictably, it turns out that 'Trotsky, as a true doctrinaire, was insensitive to everything that was happening around him' (III, 213), and that Lukàcs' ambition was to produce a justification for 'the typical Communist contempt for facts'. 'Lucàcs' achievement', we are told, 'is to have elevated the practice of contempt for facts as compared with "systems" to the status of a great theoretical principle' (III, 304). There were of course attempts by the 'so-called New Left' to mollify the savage conclusions to which Marxism led; but Kolakowski reports that 'discussion made it increasingly clear that Stalinism was the natural and legitimate continuation of Lenin's ideas' (III, 460). Lukacs had at least been 'consistent and perfectly clear' in arguing that 'the Communist Party is infallible'; critical theory, however, fudged the issue in 'an inconsistent attempt to preserve Marxism without recognizing the class or party criteria of truth' (III, 355, 357). So the last volume closes with Marxism brought to its tragic completion — annihilated by 'the soteriological strain which was blurred in Marx himself', tripped up by its own 'neo-Platonic roots', and unable to escape 'the link between Marx's belief in the prospect of man's complete reconciliation with himself, and the neo-Platonic gnostic tradition that found its way into Marxism through Hegel' (III, 448). So Marxism was really no more than a perverted religion, mesmerized by a hubristic, narcissistic 'self-deification of mankind'. It presented itself as a super-science, but it neglected the need for laborious, sceptical empirical inquiry; so in the end it revealed itself — in the words of Kolakowski's concluding epitaph — as 'the farcical aspect of human bondage' (III, 530).

The Miraculous Locksmith

As a biographer of Marxism, Kolakowski can be criticized both for his poor literary style (badly translated too, and carelessly proofread), and for his clumsy technique as a story-teller. He has no sense of suspense: unabashedly, he secretes ends in beginnings, producing a claustrophobic feeling that futures have never been open, and ought not to have seemed so. He is the omniscient author, who never allows his characters a plausible and independent point of view: each event in his tale is absorbed into its distant effects. Kolakowski tells us, for instance, that he interprets the young Marx in the light of the old so that he will be able to present what he calls the 'beginnings' or 'germ' or 'obscure prefiguration' of the old in the young (I, 100–104). And he says that even if Leninism was not predestined for dominance in Russian Marxism, 'it is hard to say that the perspective created by later events, and by what we know today of the historical consequences of Russian Marxism, is a false one'. Attempting to excuse his rather crude recourses to wisdom after-the-event, Kolakowski observes that it is impossible to describe a historical event 'as if we knew no more of its effects than people did at the time' (II, 355).

Consequently, Kolakowski sets his thinkers in a deadening framework of hindsights in which they are never seen as confronting or investigating the world or as appraising, interpreting, criticizing or perhaps misunderstanding the thoughts of others, or as being amazed, anxious or ambivalent about them. He tells a story, instead, of 'influences' transmitted across the centuries, rather like the bacilli of a great plague, their nature unaltered by their hapless hosts. So, for example, Kolakowski's description of Sorel proceeds as follows:

> His Jansenist upbringing no doubt gave him a dislike of any optimistic faith in the natural goodness of mankind . . . From the same source came his contempt for Jesuitical tactics of conciliation . . . His technical education and work as an engineer instilled into him a cult of expertise and efficiency. . . . From Marx he learnt to believe that the revolution that would restore society was to be carried out by the proletariat . . . Giambattista Vico contributed the notion of *ricorso*, the cyclical return of mankind to its own forgotten sources . . . The importance Sorel attaches to family and sexual morality in social life is due to Proudhon . . . In addition, Sorel derived from Bergson a conviction of the inexpressibility of the concrete. . . . The influence of Nietzsche is clearly felt in Sorel's cult of greatness . . . The great exponents of liberal conservatism — Tocqueville, Taine, Renan — exercised a strong influence . . . (II, 115-52).

It sounds more like a charity bring-and-buy than an intellectual tradition. It is hard to see what could be the point, in a world like this, of attempting to think for oneself; for one would, it seems, be predestined to end up as nothing better than a heteronomous permutation of other people's 'influences'.

Another device which Kolakowski uses to seal off his narrative is his repetitive description of unorthodox thinkers as 'not Marxist'. To Kolakowski, for example, 'it does not seem that Benjamin had much in common with Marxism, despite his occasional professions'; 'his work is not important as a contribution to the development of Marxism'; and 'the work . . . cannot be wholly summed up in terms of the history of Marxism' (III, 350, 348, 344). To be a Marxist, according to Kolakowski, it is essential to be 'orthodox'. An 'orthodox Marxist', in Kolakowski's book, is not someone who interprets Marx correctly and agrees with him, but someone who believes, in advance, that all and only what Marx wrote is true: the miraculous locksmith. The main stream of Marxism, thus, is the work of 'professional Marxists . . . imbued with "orthodoxy" in the sense that, whatever they are engaged in, they never forget that the purpose of all their endeavours is to defend and exalt the doctrine of which they are custodians' (II, 193). It follows, of course, that Marxism has had a debilitating effect on rational inquiry: it would not have been 'orthodox' otherwise. Kolakowski's final judgment on Marxism is that it ended by losing 'its clear-cut doctrinal form: it became merely one of several contributions to intellectual history' — not a wholly ignominious fate, one might think — 'instead of an all-embracing system of authoritative truths among which, if one looked hard enough, one could find the answer to everything' (III, 465).

As a Reference-Book

It is time I stopped treating this book as if it were a pure literary icon, with no extratextual references. Few will read it from beginning to end, so as to grasp its overall narrative design. Most will turn to it as a work of reference and will be particularly grateful for its unrivalled account of the history of Marxism in Poland. But even as a reference-book it has its disadvantages. The indexes (one for each volume) are inadequate as a guide to concepts and arguments, and the bibliographies are out-of-date, inconsistent and unreliable. Moreover, the scope is far from comprehensive: Japan, India, North and South America and the whole of Africa are wholly ignored, without explanation.

Britain scarcely fares better, receiving only three brief discussions. For the period of the Second International, we are informed that 'British Socialism was hardly affected by Marxist doctrine'. This is a rather rash dismissal of organizations like the SDF, the SL and the SLP, not to mention the presumably 'Marxist' doctrine of the Fabians, who believed that capitalism would be forced by its own irrationalities to yield to socialism, just as feudalism had earlier been replaced by capitalism. Kolakowski attempts to fit the facts to his fable by saying that '*Fabian Essays in Socialism* (1899), which struck the keynote of British socialism for generations to come, comprised a programme of reform which was either contrary to Marxist theory or rooted in principles drawn from the general arsenal of nineteenth-century socialism'. Perhaps it is trivial to complain that Kolakowski gets the date wrong (it should be 1890): but it is not mere pedantry to insist that the *Fabian Essays* — especially the contributions of Shaw and Webb — have affiliations with Marxism which call for serious investigation, or to object that on Kolakowski's definitions it seems to be self-contradictory to call an idea Marxist if it is also 'rooted in . . . nineteenth-century socialism'. Again, an investigation of socialist reading in Britain is unlikely to support the idea that *Fabian Essays* 'struck the keynote of British socialism'. Undeterred, Kolakowski draws out his preset conclusion, that Britain 'made no significant contribution at this time to the evolution of Marxist doctrine . . .' (II, 14-15).

As for British Marxism in the 1920s and 1930s, there is only a vacant reference to Haldane, Dobb, Laski, Strachey and Caudwell, and a bouquet for George Orwell, who 'formed an idea of Communism in action from empirical facts instead of from doctrinaire assumptions' and who was therefore 'met with hatred and indignation'. Kolakowski, himself not overburdened with 'empirical facts' about British Marxism, concludes that 'hypocrisy and self-delusion had become the permanent climate of the intellectual left' (III, 114-16).

Post-war British Marxism, for Kolakowski, is dominated by 'the New Left', which found expression in the *New Reasoner* and the '*University and Left Review*' (*sic.*). The New Left bred the 'ideological fantasies' which, given expression in 1968, revealed themselves to be 'no more than the nonsensical expression of the whims of spoilt middle-class children ... They long desperately for a miracle, they believe that a single magic key will open the door to paradise, they indulge chiliastic and apocalyptic hopes ...' (III, 488-90; cf. III, 180). By suppressing all reference to the political events to which the 'New Left' was responding — Macmillanite Toryism and Gaitskellite socialism at home, Algeria, South Africa, Vietnam, Berlin, the Soviet Union, and indeed Poland — Kolakowski adds a specious verisimilitude to his dismissal of its ideas as 'fantasies'. However, when you consider the long list of 'New Left' intellectuals — Thompson, Hill, Anderson, Hobsbawm, Sedgwick, Meek and dozens of others — it is clear that he is in no position to sniff disdainfully at slovenly scholarship. As a general reference-book Kolakowski's work will not stand comparison with an old classic like G.D.H. Cole's *History of Socialist Thought* and is for most purposes superseded by Tom Bottomore's *Dictionary of Marxist Thought*.

TWO: SCIENCE AND PHILOSOPHY IN THE UNDERSTANDING OF HISTORY

'Karl Marx was a German philosopher': so begins the first volume of *Main Currents*. 'It is the purpose of this inquiry to understand Marx's basic thoughts as answers to questions that have long exercised the minds of philosophers... Marxism can be described as constituting a new step in the development of European philosophy' (I, 1, 11). Philosophy, to Kolakowski, is not to be confused with historiography, economics, politics or social theory. It is a wholly speculative, non-empirical inquiry into rival definitions of truth, freedom and goodness. Kolakowski relies on the traditional vocabulary and procedures of the historian of philosophy in order to pigeonhole the doctrines of his protagonists — an approach which, notoriously, tends to belittle the labour of serious philosophical thought, rendering it down to the whims of fashion, with philosophers simply choosing whatever off-the-peg garments they fancy, for unreal contests where nothing is really at stake anyway. So in Kolakowski's pages we see Marx disputing epistemology with Descartes and Kant (I, 134), from a position of 'generic subjectivism' (I, 176). Engels preferred to be 'a radical empiricist as regards the genesis of knowledge . . . and a moderate empiricist as far as method is concerned . . .' (I, 395). Then Kautsky appears with 'a naturalistic and positivistic view of consciousness' (II, 42); Jaurès, however, opted for 'a universal soteriology of Being' (II, 125), whilst Brzozoski took 'biological relativism' (II, 221), Mach and Avenarius went for 'a form of scientism and positivism (II, 431), and 'Gramsci's position is one of species subjectivism and historical relativism' (III, 249). As usual with historians of philosophy, we are given a history of thought in which no one seems to have to do any thinking: they just 'take positions' and sit tight.

When he returns to Marx, Kolakowski asserts that 'the essential feature of Marxism consists in the doctrine that the act of understanding the world and the act of transforming it achieve identity in the privileged situation of the proletariat' (III, 393). The key terms of Marx's theory, for Kolakowski, are 'alienation', 'reification' and 'the philosophy of *praxis*' (I, 104, 277; II, 3, 61; etc.). This interpretation is open to some well-known objections — such as that Marx himself omitted to use such terms as 'philosophy of *praxis*' and 'reification', and that he saw himself as contributing to the understanding of economic theories of surplus-value, rather than to philosophical theories of freedom and truth. Kolakowski acknowledges none of these difficulties, however, but shelters behind the authority of Lukacs — who, paradoxically, appears elsewhere not as the creator of the 'correct interpretation of Marx's philosophy' but as an unprincipled apologist of Communist 'contempt for facts' (I, vi; II, 297).

Philosophers have a habit, disconcerting for non-philosophers, of vehement disloyalty to their subject — proclaiming that philosophy, as distinct from science, is a tip of vain errors, and that the philosopher's task is to warn others to keep off. It is thus without any trace of solidarity that Kolakowski treats Marx as a fellow-philosopher. When he qualifies Marx's conception of the proletariat as 'a philosophical deduction', for example (I, 130), he means to bury Marxism, not to praise it. And he holds that compared with what he calls 'science' and 'scientific opinion', philosophy is no better than 'ungrounded prophecy' (I, 373). Hence, on Kolakowski's scheme, it follows from the fact that 'Karl Marx was a philosopher' that Marxism is 'not a scientist's theory but the exhortation of a prophet' (I, 375).

One of philosophy's vices, according to Kolakowski, is its

tendency to ignore, or deny, the gulf between 'facts' and 'values'. The idea that there is such a gulf, and that reminders of its existence are all that can validly issue from a modern moral philosophy, has of course been a fervent belief amongst philosophers of Kolakowski's generation, whether logical positivists or existentialists. But the fact/value dichotomy has never received any satisfactorily precise formulation, and in the past twenty years it has indeed been abandoned or smothered in qualifications by nearly every moral philosopher except Kolakowski. It is curious, then, to see how imperturbably he presents his own opinion on the matter as if it were unassailable fact, requiring neither explanation nor defence. He then excoriates the Lukacsian 'philosophy of *praxis*' (which he has identified with Marxism) for its neglect of it — 'the absence of this distinction', as he puts it, being 'fundamental to Marxism' (II, 254; cf. III, 298). 'A value judgment', he asserts, 'cannot follow from any mere descriptive analyis' (II, 360; cf. III, 286); and those who try to mingle them, he pronounces solemnly, suffer from 'an incurable contradiction' (III, 323).

Hegel, Marx and the Positivity of Error

Readers of Kolakowski's earlier works will not be surprised by the bullying and shrill dogmatism with which the positions of *Main Currents* are asserted. Kolakowski appears even to have lost the vestigial capacity for irony displayed in his essay 'My Correct Views on Everything: A Rejoinder to Edward Thompson'. What one might not be prepared for is the fact that in this massive cautionary tale about Marxism — the 'philosophy' which masqueraded as a science — the most plausible account of the relations of philosophy and science within Marx's work simply does not get mentioned. Kolakowski's oversight comes about as follows.

In spite of his repeated professions of scrupulous circumspection and meticulous scepticism, Kolakowski places enormous trust in that slipperiest category of philosophical thought — 'the dialectic'. He writes as if a contraption called 'the dialectic' had been fashioned by early Christian 'soteriology' and then passed on from generation to generation in the history of philosophy like a baton in a relay-race. 'Dialectic' is taken to be the relationship between the Absolute and the Contingent; its origins, for Kolakowski, are in Plotinus; in Hegel's hot hand, it became the story of how 'consciousness overcomes its own contingency' (I, 70); and then in Marx's, 'the dialectic . . . is the unity of theory and practice' (I, 400) or, for that matter, 'the dialectic is the consciousness of the working class' (I, 323).

These attributions — which have no basis in Marx's (mercifully few) discussions of 'dialectic' — help Kolakowski to present Marx as making a contribution to European philosophy by carrying on the tradition of Plotinus. But they also enable him to pass over the themes which were hitched on to the discussion of philosophy, science and 'dialectic' by the one philosopher Marx found it impossible to settle his accounts with: Hegel.

For Hegel, every human situation has to be seen from two points of view: First, there is the 'for-consciousness', or how things seem to those actually involved: And then there is the 'in-itself', or how things would seem to impartial and omniscient observers, able to see the meanings of actions in a way that participants never can. According to Hegel, every philosophy is an attempt, laudable but perhaps necessarily unsuccessful, to achieve the point of view of the in-itself, for which the ways things present themselves 'for-consciousness' will always be illusions. Illusions, however, are not merely negative. (If people orient themselves by reference to rainbows, and organize quests for the crock of gold at their end, then their behaviour will have to be explained by reference to the locations which rainbows have for them ('for-consciousness') — even though in reality ('in-itself') rainbows are mere appearances, with no true location at all.) Illusions, for Hegel, are the immediate motive of all human activities. Even theoretical reflection must start with illusions, so they are the ground of wisdom too. So progress arises not from shunning error but from dwelling in its midst and trying to understand and modify it. Hence the Hegelian idea of 'criticism' and of what one might call 'the positivity of error'.

Famously, Marx was impressed by Feuerbach's use of this idea to explain Christianity. For Feuerbach, God was an illusion, but not a mere illusion — an attribution of human realities to a non-existent subject, and a misperception, therefore, of what was really true. You could not undertand religion without understanding God, and you could not move beyond it — to atheism and humanism — except by standing, as it were, on God's shoulders. The significance of this theme for Marx's project of a 'critique of political economy' has been ably advocated in recent years by many writers (Lucio Colletti, for example). Marx can be seen as proposing, in effect, that 'commodity production' — that is, the situation where goods are produced for exchange in an anonymous 'market' governed by the Law of Value — is another case requiring analysis in terms of how things appear 'for-consciousness'. The mechanisms of such an economy can then be interpreted as illusions in the positive, Hegelian sense; and the concept of Value can be taken not as representing the way things really are (the 'in-itself'), but as indicating the starting-point for a better knowledge of human history than orthodox, classical economics provides.

According to this interpretation of Marx's use of Hegel, Marx's discussion of Value can be understood as follows: 'Value', he takes it, is the name for the fundamental category around which 'commodity production' revolves: it is to be considered equalized when two sets of commodities are fairly exchanged. Economists before Marx agreed that the measure of this quantity was 'labour' — roughly, that two things had the same value if it took the same amount of time to make them. This was the traditional Labour Theory of Value. Marx went further, and said that Value was nothing more than this measure — in other words, that it had no source or 'substance' other than labour. That is Marx's Labour Theory of Value. And Marx interpreted capitalism as a specially developed form of commodity-production. It was a movement whose tendency was to treat everything — religion, pleasure, art, education and, most crucially, labour itself — as a commodity, fairly exchangeable for any other commodity of the same value. This development can be called the extension of the jurisdiction of the Law of Value.

So Marx's concept of Value functions quite like Fuerbach's concept of God. God and Value do not have any reality 'in themselves'; they are passing illusions, not eternal verities. Just as God belongs to the 'for-consciousness' of a religious society, not to its 'in-itself', so it is with Value. It is not a real independent force, but those involved in commodity production subject themselves to its Law (the Law of Value) just as if it were.

No doubt Marx's theory, or Feuerbach's, has its obscurities. What can it mean, for example, to say that the 'substance' of Value is *really* labour, or that God is *really* human? And in Marx's case, the theory is complicated by his conception of 'exchange value' as the 'form of appearance' of Value itself. Both ought to be recognized, however, as ingenious atempts to use a Hegelian framework in order to analyse commodity-producing (or God-worshipping) societies as having a passing but important significance in human progress considered as a whole — even if, in the end, one is forced to the now-fashionable conclusion that such comprehensive theoretical ambitions are bound to be disappointed.

Philosophy, Science and the Theory of Value

One effect of Marx's, or Feuerbach's, adaptation of the Hegelian view of progress is that it complicates the relations between 'philosophical' and 'empirical/scientific' components in their views of history. It means that one cannot assume, with Kolakowski, that the philosophical and empirical elements must be antithetical to each other, or inversely proportional. For it might be that, without philosophical concepts, some empirical facts could not be scientifically recorded.

Brashly, Kolakowski attempts to refute Marx's concept of Value without even noticing the possibility of this interanimation of philosophy and science within Marx's view of history. He takes it that Marx intends Value to be taken as a fundamental, ahistorical, explanatory concept — comparable, perhaps, to Attraction in Newton. Then he complains that Value is not precisely measurable, which according to his hasty inference means that 'it cannot be applied in the empirical description of phenomena'. (This, according to the interpretation neglected by Kolakowski, is like saying that an anthropologist studying a religion may not use concepts like 'God' or 'afterlife' unless he can prove their reality and show how they are to be measured.) Pursuing his 'refutation', Kolakowski repeatedly commits an elementary confusion — between the Labour Theory of Value (i.e. the definition of Value in terms of labour) and the Law of Value (i.e. the rule to which a society is subjected to the extent that it is given over to commodity-production). Kolakowski then subjects this composite fudge to what he sees as basic tests of scientific health and, not surprisingly, pronounces it totally unfit. 'It is', he says, 'an ideological, not a scientific category, and cannot be verified empirically' (II, 296). A natural scientific law, says Kolakowski (presumably thinking of physics), 'is generally a statement that certain phenomena occur in certain circumstances; but it is not clear that Marx's definition of value can be expressed as a law. Marx's theory . . . that labour-time . . . is the only constituent of value . . . is not a law, but an arbitrary definition which cannot be proved empirically and is of no use for the empirical description of economic phenomena . . .' (I, 326–27).

Having confused the Law of Value with the Theory of Value, Kolakowski inevitably overlooks the fact that Marx was putting forward the theory (surely empirical) that certain societies actually do tend to regulate their affairs by a principle (the Law of Value) which depends on a category — Value — defined by the Theory of Value. In a saddening outburst, Kolakowski jeers at Marx for unjustifiably postulating that the ploughman's labour creates Value, whereas that of his horse does not (I, 329–30). But the arbitrariness is not Marx's: he was merely trying to specify the rationale behind the fact that in developed commodity-producing societies it is normal to give wages to the ploughman; whereas, as Kolakoski ought surely to concede, it is regarded as unnecessary to give wages to the horse. So, on this reading, it is exactly *because* of its 'philosophical' character (not its borrowing of something called 'the dialectic' but its adaptation of the Hegelian contrast between the 'in-itself' and the 'for-consciousness') that Marx's view of history has empirical or 'scientific' implications; and so Kolakowski's fulminations about 'science' and 'philosophy' rebound upon their author.

Kolakowski's account of Marxism's ambitions as a superscience has been criticized by some Western Marxists for failing to observe a Great Divide between an authentic, liberatory Marxism, based on a philosophical vision of Communism, and the unphilosophical, deterministic and oppressive practices of 'Stalinism'. One thing that Kolakowski's readers must be grateful for is the reminder that this distinction is fanciful — for Stalin's policies, and Mao's, were themselves devised and defended in terms of large philosophical notions, of Dialectic, Liberty and Truth. The Great Divide, in other words, is a myth, and it has lulled Western Marxists into a wholly unwarranted complacency when confronted by the unacceptable face of Marxism elsewhere in the world. Kolakowski's critics cannot cling to the idea that a 'philosophical' Marxism can be separated from a 'scientific' variety, so that the former is left with no blood on its hands. But the mistake of both Kolakowski and his critics is their view that genuine Marxism is 'really' philosophical, and that therefore it cannot be 'scientific' or 'empirical'.

THREE: THEORETICAL UTOPIA

As a professional philosopher, Kolakowski naturally pays homage to intellectual 'clarity'. In pronouncing his negative verdict on nearly every thinker he discusses, his refrain is always: 'he does not make it clear . . .' The distinction between productive and unproductive labour 'is highly obscure', for instance, and 'its purpose is not very clear'. 'It is also not clear', he goes on, why Marx says that wages represent labour power rather than actual labour, and 'it is not clear how Marx's assertion helps towards an understanding of present-day capitalism' (I, 331–32).

Kolakowski's own account of truth and ambiguity, however, is itself unclear, or even incoherent: in an impatient aside, he says that since rain is sometimes beneficial, sometimes not, 'the statement "rain is beneficial" is ambiguous'. (If you think about it for a second, you will realize that it is because the statement is *not* ambiguous that it is true of some situations and not of others.) Astonishingly, Kolakowski thinks that the idea 'that a statement, without changing its meaning, may be true or false according to circumstances... belongs to the category of nonsense'; and this, he concludes with evident pleasure, is something which 'we did not... need Marx's intellect to discover' (III, 155–56).

But even if a virulent 'obscurity' does disfigure the basic theses of Marxism (and, for all his vehemence Kolakowski is far from proving that it does), this will not justify a comprehensive dismissal of them all. There are three reasons. The first is that such ambiguities should not simply be noted, as fatal follies, whose existence is evident to trained philosophical inspectors like Kolakowski. Their nature, motivation and function ought to be specified, and attempts should be made to remove them. (Otherwise the suspicion will become confirmed that philosophy is the last refuge of intellectual sluggards, protecting their sceptred isle of ignorance by sneering at any substantial body of theory which appears on the horizon of their discipline, rather than joining in understanding, criticizing and improving it.)

The second objection to Kolakowski's obsession with philosophical clarity is that it stops him seeing that a theorist may be well-advised to leave some questions open. The indisputable fact, for example, that Marxist statements about the 'economic base' and its cultural-political 'superstructure' are liable to have an 'on the one hand, on the other hand' character — as Kolakowski points out more than once (III, 63, 104) — means that they are rough concepts, but does not prove that they are useless. Moreover, a controlled keeping-open of a question may be a virtue in a piece of writing, and testimony to special deftness. Several of George Meredith's novels carry off the remarkable feat of exhibiting a character actuated by a thought which neither the narrator nor anyone in the story — even the person who has it — can identify. It would be fatuous, though, to criticize Meredith for being 'unclear', or to try to sort out the 'ambiguity'. Marx's writings perhaps resemble Meredith's in that respect; and certainly Hegel's do. However, Kolakowski advises us that 'despite all the ambiguities of the *Phenomenology of Mind* ... Hegel's metaphysical epic affords sufficient clues as to its general intentions'. He then 'improves' it into a crass and authoritarian sketch of the Progress of the Race, though, even then, he finds that it is 'beset with ambiguity' (I, 68, 79).

The third objection to Kolakowski's attitude to theoretical 'ambiguity' is that it may set impossibly high standards. No doubt obscurity and contradiction are undesirable; but what if there is no available vocabulary which is free of them? If you set your heart on instant conceptual perfection you may just get intellectual paralysis. If the history of physics, to which Kolakowski often defers, has any general moral, it is surely that paradigmatically 'scientific' theories teem with ambiguity, obscurity, and even contradiction, but may still provide knowledge and a degree of technical control. Kolakowskian squeamishness about 'unclear thoughts' is rather like refusing to live in a house which isn't guaranteed to stay up for ever. He often reproaches Marxists for their Utopianism — for neglecting the flawed social achievements of capitalism while yearning for the never-never land of socialism; but perhaps it is the anti-Marxists like Kolakowski, who dream of a social theory of 'absolute clarity', who are the really deluded Utopians.

'The Communist Mentality'

The defects of Kolakowski's narrowly philosophical approach to the history of Marxism carry another one in their wake: Kolakowski charts the theoretical odyssey of Marxism simply in terms of its expression in published books. He makes no reference to the ways in which — for whole classes and whole countries, as well as for millions of scattered individuals who do not write books and may not even read them — Marxism has opened up possibilities of adventurous if imperfect thought about unrealized social hopes; albeit on the basis, perhaps, of coarse slogans (such as 'unity of theory and practice'), wrongly thought to have Marx's authority behind them. The insights based on these may not entitle one to a chair of Philosophy in an ancient University, but they might be subtle and effective knowledge nevertheless. The important question is whether thay are preferable to any real alternatives.

Kolakowski allows this dimension of popular belief into his story only once, when he tries to explain why 'the analytical standard of Marxism philosophy' (*sic*) has been 'so low' in the Soviet Union. His explanation is built on the concept of 'Stalinism'. Stalinism, Kolakowski thinks, is the Satanic creation of Communist Parties, East and West. It is a product of 'the Communist mentality' — 'a mentality that was completely immune to all facts and arguments "from outside", i.e. from "bourgeois" sources' (III, 452). It is hard to see how Kolakowski would explain why millions of people (including his former self) have thought that there was something worthwhile in a 'mentality' so vacuous and vicious, let alone how a sainted few (such as himself) managed to think their way out of it again.

He suggests that Russians are bad at logic because 'Russia did not go through a scholastic phase', and that Russians are unschooled in 'scepticism and relativism' because 'Russia also had no part in Renaissance civilisation' (II, 308); a view which depends on a rather negative attitude towards non-Western European cultures and neglects the fact that most of the Communist world, or indeed of the Soviet Union, is not Russian. His other, more general, explanation relies on a notion of 'the parvenu mentality' — 'the mentality, beliefs and tastes of someone enjoying power for the first time'. Somewhat incautiously for a person who offers himself as a champion of 'purely intellectual criteria of empiricism or logic' (III, 340), Kolakowski ignores the fact that power in the Soviet Union tends to be concentrated in the hands of a hereditary educated elite. He affirms that the Soviet State is run by

> individuals of worker and peasant origin, very poorly educated and with no cultural background, a thirst for privilege and filled with hatred and envy towards genuine "hereditary" intellectuals. . . . A parvenu has no peace of mind as long as he sees about him representatives of the intellectual culture of the former privileged classes . . . he endeavours to convince himself and everyone else that his native country or milieu is superior to all others. . . . The parvenu combines a peasant-like subservience to authority with an over-mastering desire to share in it . . .

Hence, according to Professor Kolakowski, the 'mania' and 'megalomania' of 'Stalinism', and the 'grotesque' and 'monstrous' forms of the 'cult of the Leader'; and hence, too, 'Stalin's philosophy', which 'was admirably suited to the parvenu bureaucratic mentality in both form and content' (III, 149, 151).

It would be harsh to deny to someone with Kolakowski's experience of repression and exile the right to passionate denunciation of the cultural policies of the Soviet Union. But he ought not to falsify the record — for instance, by describing the Proletkult movement as an expression of 'hatred of education' (III, 514). Nor should he equate social processes as different as those of China under Mao and of the Soviet Union under Stalin. Remarkably, Kolakowski claims that Maoism — which in fact involved mass literacy campaigns — was based on 'the idea that illiterates are naturally superior to scholars'; and he adds that its 'deep mistrust of learning, professionalism, and the whole culture created by the privileged classes illustrates clearly the peasant origins of Chinese Communism', which he sees as just another example of 'the traditional hatred of peasants for an elite culture' (III, 513–14).

The use of such mass-psychologisms in political debate is, however, liable to recoil on the author. If the 'Communist mentality' is a result of peasant megalomania, perhaps Kolakowski's response could simply be attributed to aristocratic paranoia. Or if it relates to centuries of Oriental backwardness, then perhaps Kolakowski's answer is simply a continuation of centuries of envious identifications with 'Western culture' by French- and English-speaking Polish intellectuals. This brings me back to the question of the authorial personality which presides over *Main Currents of Marxism*.

Above It All

These volumes are not so self-effacing as they appear. True, Kolakowski's own part in the development of Marxism is passed over with a becoming modesty. But he indulges in abusive sarcasm towards his caricatured Marxist victims. For instance he alleges that any serious Marxist theory of art will have to imply that 'one could write the works of Shakespeare if one knew enough about the economy of Elizabethan England' (II, 345; see also II, 431, 487, 506, 513; III, 3-4, 44, 61, 62, 85, 172, 182, 237, 284, 364-5, 417, 422, 435-6). All this leaves the reader wondering how Kolakowski proposes to leap out of this sizzling pan of error and what he expects to land in when he does. He seldom acknowledges that the problems which Marxism has run into have been objective ones, which would confront people even if Marxism had never been invented. For instance, he sneers at Lunacharsky because 'like most Marxist theoreticians on art, at all events educated ones — he had difficulty in accommodating his "bourgeois" tastes to his "proletarian" ideology' (II, 445) — as if the 'difficulty' about the relations of democracy and high culture was a pure phantom of the Marxist's brain!

The author of *Main Currents* presents himself as immaculately above them all, as having been through them and left them behind. Like an old-fashioned psychoanalyst, he accuses those who reject his interpretations of refusing to grow up. Such wisdom as his, we are given to understand, requires a mature and clear-sighted moral courage which no Marxist could ever possess. The trouble with Lukacs, for instance, is that he 'craved intellectual security and could not endure the uncertainties of a sceptical or empirical outlook' (III, 306), while American students are afflicted by a 'contempt for technique and organisation' which 'goes hand in hand with a distaste for all forms of learning that are subject to regular rules of operation or that require vigorous effort, intellectual discipline, and a humble attitude towards facts and the rules of logic' (III, 420).

And so, in negative, emerges the real rhetorical accomplishment of these volumes: a portrait of a saintly author, a saddened but all-wise fugitive from apocalypse, transfigured and, now, above it all. And hence the extraordinary prestige of the book: actually a rather derivative, repetitive, unreliable, and ill-written rag-bag, but decorated with endorsements such as Sidney Hook's ('Kolakowski's magisterial study must be acknowledged as opening a new era in Marxist criticism') or Steven Lukes' ('deeply impressive... marked by lucid and accurate exposition, sustained and high-level analysis') or Geoffrey Hawthorn's ('a work of surpassing lucidity and power, of the sharpest and most sensitive judgement, of a far finer quality than most of that with which it deals . . . a masterpiece'). It is certainly hard to remain unmoved by the image of Kolakowski as a wise old man who, having wandered through all the highways and byways of world Marxism, has at last seen all its problems vanish as mere fantasy. The welcome which the book has received is not hard to explain — for who would not applaud such a figure, and even in some ways identify with him? But in fact, this sanctified image of the author is a deceptive contrivance — the most misleading and meretricious fantasy in the whole book.

References

I am grateful to Russell Keat, Les Levidow, John Mepham, Peter Osborne, Roger Owen and Bob Young for comments on drafts of this essay.

Tom Bottomore, ed., *Dictionary of Marxist Thought*, Oxford, Blackwell, 1983.

G.D.H. Cole, *History of Socialist Thought*, 5 Vols., Macmillan, 1953–60.

Lucio Colletti, 'Introduction' to *Karl Marx: Early Writings*, Harmondsworth, Penguin, 1975.

Leszek Kolakowski, *Chrétiens sans Eglise: La Conscience religieuse et le lien confessionel au XVIIe siècle*, Paris, Gallimard, 1969; from the original Polish (Warsaw, 1965).

Leszek Kolakowski, *Marxism and Beyond*, Pall Mall, 1969.

Leszek Kolakowski, 'My Correct Views on Everything', in *Socialist Register 1974*, Merlin, 1974.

Henry Pachter, 'Unfair to Marx', *Dissent*, 26, 3 (Summer 1979), 338–43.

Jonathan Rée, 'Socialist Humanism' (an account of the Thompson-Kolakowski relationship in terms of the history of the English 'old New Left'), *Radical Philosophy* 9 (Winter 1974) 33–36.

E.P.Thompson, 'An Open Letter to Leszek Kolakowski', in *Socialist Register 1973*, Merlin, 1973; reprinted in E.P. Thompson, *The Poverty of Theory and Other Essays*, Merlin, 1978.

Marx Wartofsky, 'The Unhappy Consciousness: Leszek Kolakowski's *Main Currents of Marxism*' (a sustained attack on Kolakowski's thesis that Stalinism is a natural consequence of Marxism), *Praxis International*, Vol. I, No.3 (October 1981), pp. 288–306.

SCIENTISTS AS MILITARY HUSTLERS

Bruno Vitale

How should we understand scientists' involvement in military-funded research? Critics tend to see it as naivety and/or a regretful succumbing to external pressures. Although these aspects are relevant, they obscure the decisive initiatives that scientists have often taken to propose research intended to solve military problems. Scientists should be seen as willing partners in this collaboration as they go about devising new ways to make science useful for military purposes.

A Rich 'Menu'

There seems to be widespread consensus among qualified observers that a rich technological menu has been offered to the military by the scientists and engineers (Knorr and Morgenstein, p. 14).

. . . [during World War II] the scientists, in part, defined their own problems and took the responsibility for selling not only new techniques but also strategies to go with them to the military services. Scientists appeared on the battlefield zones as civilians, flying with bombing missions to observe the operations for new developments (Yarmolinsky, p. 289).

This is a fact: scientists, within the institution of science, are always enriching the menu that they periodically propose to the military. They feel personally concerned with development and production and deployment of new weapons systems. They like to be on the spot, briefed by generals, treated as honoured guests.

The recent history of the development of the neutron bomb is unambiguous: scientists wanting the bomb, scientists lobbying for the bomb, scientists proving that the bomb would solve most military problems. The survival of a research laboratory, the prestige of a small group of scientists, leads to the development of a new instrument of mass destruction. A (heavily censored) transcript of a 1973 hearing of the US Congressional Joint Atomic Energy Commission is illuminating. H. Agnew, director of the military labs at Los Alamos, stated:

> It may be that people like to see tanks rolled over rather than just killing the occupants. It is quite clear there is rethinking going on ... I know we at Los Alamos have a small but very elite group that meets with outside people in the defence community and in the various think-tanks. They are working very aggressively, trying to influence the Departments of Defense to consider using these [deleted] weapons (*New Scientist*; also Vitale 1981/82).

Scientists take the initiative. They did so particularly well during World War II. They were at the root of the programme leading to the atomic bomb, as is clear from the famous letter signed by Einstein and addressed to F.D. Roosevelt (2 August 1939):

> This new phenomenon (the nuclear chain reaction in Uranium) would also lead to the construction of bombs, and it is conceivable — though much less certain — that extremely powerful bombs of a new type may thus be constructed. A single bomb of this type, carried by boat and exploded in a port, might well destroy the whole port together with some of the surrounding territory ... In view of this situation you may think desirable to have some permanent contact maintained between the Administration and the group of physicists working on chain reactions in America (Strauss, p. 178).

But the collaboration operated at all levels, in all fields. Here is a chilling example from a memorandum sent by a scientist (H. Ewell) to V. Bush and transmitted to the Army Air Forces:

> Advance estimates of forces required and the damage to Japanese war potential expected from incendiary bombing of Japanese cities indicate that this mode of attack may be the golden opportunity of strategic bombardment in this war — and possibly one of the outstanding opportunities in all history to do the greatest damage to the enemy for a minimum of effort (Baxter, p. 97).

I have given two examples from two outstanding successes by scientists. Atomic bombs were, of course, produced and used on Hiroshima and Nagasaki. By the end of the war, incendiary bombing accounted for more than 3,500,000 dwelling units destroyed or heavily damaged, more than 5,500,000 Japanese people rendered homeless, more than 300,000 dead, and more than 750,000 wounded (Baxter, p. 99).

I could give many more examples. But I want to emphasize that this is only part of the story; I am beginning to feel that it is the part of the story that is most suited to the military and to the ruling classes. In what follows, I shall link the 'rich menu' ideology with the 'technological inevitability' ideology. They both contain, of course, quite a large amount of truth; but, taken in isolation, they contribute to building up an image of the military as bound to the progress of technology. This is a very useful image, one that hides responsibilities and complicities. And, anyway, *who* has created this institution of science that insists on offering 'rich techno-logical menus' to the reluctant military?

Technology as a 'Mighty Rock'

> Since military technology as a whole is the most rapidly developing aspect of comtemporary military affairs, one can assume beyond doubt that it is the development of armaments which, in the last analysis, determines the development of the whole of military affairs. This is a fundamental proposition on which the entire present chapter is based . . . At the present time the rapid development of military technology has effected a still greater influence on the development of military affairs . . . Soviet science and culture form a mighty rock on which to base the solution for any problem of military affairs or of military technology (Powkrowsky, pp. 64, 101, 126).

It seems to me that the old and discredited use/abuse model of science in our society is being replaced smoothly by a more subtle model, that integrates the 'rich menu' ideology already presented, and the 'technological inevitability' ideology exemplified by the quotation above. Shall we call it the 'military clean hands' model?

The use/abuse model of the functioning of scientific institutions is too well known and worn too thin by age to deserve much analysis. More or less, it goes like this: Scientists are in quest of truth, being somewhat convinced that any truth about nature is of general interest to humanity. The institution of science (be it civil or military) provides them with funds and tools for their search. The results are common patrimony of man: If someone decides to use some of these results for evil, too bad; the scientists accept no blame.

You could think this as a caricature of a serious model. But listen:

> If I produced butcher's knives, I would feel totally in peace with my conscience, even if those knives were sometimes used to kill people; all said, people need butcher's knives, and it is not the producer of knives who has to concern himself with the criminal use that someone could make of them (Neel, as quoted by Aigrain).

It is a Nobel Prize winner for physics who is speaking; a long-term collaborator with the military. The caricature is not in my presentation of the model, but in the use/abuse model itself.

In comparison, the 'military clean hands' model is more subtle, leading us to believe that the military itself has clean hands with respect to science. That is, the model attributes weaponry development to the efforts of scientists-hustlers and to the inevitability of technological progress. This rationalization can hold some credibility, as we are all sensitive to the great impetus of scientific research in our world; we feel that there is an internal logic in it (hard to define, even for those who work inside science) and that this logic leads to technological development, production, application, consumption of newer and newer industrial goods and, of late, agronomic products. We feel the thrust of this impetus in all aspects of our life: we have not *asked* for colour television or for supersonic aircraft, neutron bombs, electronic printing, teaching by computer... We have been given presents (in the form of a continuous technological progress) but we have no way to refuse or discuss them.

Capitalistic production shares the same belief in the inevitability of technology: all that can be made, should be made; otherwise, a competitor will make it and destroy you. And the military seem to share the same belief: all weapons systems that can be made, should be made: otherwise, an enemy will make it and destroy you.

At this point a natural curiosity arises: what made the present institution of science able and eager to offer more and more technological enjoyment to our everyday life, to industrial production, to agricultural production, to the military? Is there an inherent perversity in knowledge which makes the search for novelty so stimulating and exciting that all considerations on risks and mass destruction and violence on a world scale are forgotten?

It is hard to believe it, as it is hard to believe that industrial production and military affairs are *victims* of the inevitable progress of technology. It is much easier to convince oneself that contemporary science and technology have been shaped by the ruling interests of our society in such a way as to be ready to *serve*. It is in this context that I prefer to talk about the institution of science rather than about science and/or technology. Science sounds a-temporal: a growing body of knowledge about natural phenomena, an increasingly complex kit of tools for the control of nature. You can talk about Greek science, or Chinese science, and you will forget the interests that shaped them. In the same way, technology sounds homely: new gadgets for our everyday life, new hopes for a longer life expectancy.

The institution of science is the whole body of the present practice of science: the scientists, the scientific/academic institutions, the funding agencies, the scientific journals and associations, the administrative personnel (including the girl who 'typed and retyped with unfailing effectiveness my unreadable manuscript . . .'), the necessary constellation of students and young research workers. All of this is needed to do science today; and it is not given free to scientists for amusing themselves. It is part of a societal project, satisfying needs and interests (sometimes contradictory), reflecting — in its structure, functioning, ideology — the larger world that makes its existence possible.

Before doing a detailed analysis of the role of military-funded research, one should unravel as much as possible these different components. It is misguided to focus on military research and military-funded research (meaning the whole set of research that — independently of their direct, indirect or null relevance to weapon systems and explicit military affairs — is funded by or through military agencies). It is misguided to try and understand their role in a vacuum, as a degeneracy of the system, as a perversion of science. I would contend that military-funded research is an integral part, a very organic and conscious part, of the whole research effort; there is some division of roles, there are different priorities posed for research, there could be different styles in research policies, but the different pieces fall into a reasonably coherent pattern.

Rather than dwell on this general picture, here I shall concentrate on military-funded research, its role, its impact on the institution of science. And — even with this qualification — I shall deal with only a small part of the whole pattern: mostly with scientists, what they are, what they think they are, what other people think they are. And, mostly, *why* they are what they are — willing to serve the powers that be.

There are several myths that should be dispelled if one wants to tackle this problem. One of the most persistent ones — together with that of the use/abuse model, which is quite consistent with it — is the following: scientists are (or could be) the critical consciousness of our society. Let us see how much there is to this myth.

Scientists as the Critical Consciousness of Society

> We, representatives of German science and art, protest — in front of
> the entire civilized world — against the lies and calumnies by means of
> which our enemies try to dirty the pure cause of Germany, in the
> difficult struggle for existence that has been imposed to her . . .
> It is not true that we have criminally violated Belgian neutrality . . . : we
> would have destroyed ourselves if we had not taken the initiative.
> It is not true that our soldiers have taken away the life or the property of
> a single Belgian citizen, except when constrained and against their
> will . . .
> It is not true that our troops have brutally ravaged Louvain. They have
> been obliged to retaliate, against frenzied inhabitants who have
> treacherously attacked them; it has been with a heavy heart that they
> have shelled the town . . . The famous Hôtel de Ville stays intact: risking
> their lives, our soldiers have kept it from burning . . .
> Without German militarism, German culture would long since have
> disappeared from the world (Kellerman; Schröder-Gudehus).

You are not dreaming and it is not a text from a book of science-
fiction: it is only a small part of an 'Appeal to the Civilized World'
signed, on 4 October 1914, by 93 famous German scientists and
artists. As we are talking about science, let us have a look at the
scientists' names, the very cream of German intelligentsia, Nobel
Prizes and the like: P. Ehrlich, E. Haeckel, F. Klein, W. Nernst, W.
Ostwald, M. Planck, W. Roentgen, W. Wien . . .

Each one of these scientists had given a fundamental contribution to his research field, by challenging established dogmas, by
painstakingly trying to restructure on a new level a domain torn by
contradictions, by looking intelligently to new facts and phenomena. And then they sign their 'Appeal': without any proof, or
possibility of checking and proving/disproving what they are
saying, or nuance about 'the pure cause of Germany'. They have
been asked to serve, and they do so, servilely. That is why I say:
they (we) have been trained to serve, and in a much subtler way than
poor soldiers in an army. What is striking about the German
Appeal is less its content than its early date: 1914, a time in which
the complex web of military agencies, military-funded research,
research laboratories, expert panels, think-tanks . . . had not yet
been woven. It was during and just after World War I that the
institution of science began to be shaped in such a way as to be
ready to provide support and to propose improvements to the
military.

As regards industrial production, this dependence already has a
long history. 'During the first three decades of the twentieth
century, therefore, the corporate engineers undertook to organize

and harness science to industry,' writes David Noble (p. 112). He identifies three phases, all of them later followed by the military: the establishment of organized research laboratories within the industrial corporation; the active support to and collaboration with research agencies outside the corporation; the national coordination of research activity in support of corporate industry. 'The research laboratories, above all, gave to the corporations command over the flow of scientific investigation. In the nineteenth century, scientific ideas had given rise to industrial manufacture; now the industrial corporations undertook to manufacture scientific ideas' (Noble, p. 118).

In the United States, out of the Civil War came the Naval Observatory, born as a 'depot of charts and instruments' and now fully grown to be the Office of Naval Research. Out of World War I came the National Research Council: '. . . the wartime NRC became a central directing agency for American science to a degree unprecedented in earlier history,' states Yarmolinsky in his perceptive analysis of the military establishment (p. 287). Out of World War II came the whole panoply of 'directing agencies for American science': the National Defense Research Committee, to 'correlate and support scientific research on the mechanisms and devices of warfare' (Baxter, p. 451); the Office of Scientific Research and Development, 'for the purpose of assuring adequate provision for research on scientific and medical problems relating to national defense' (Baxter, p. 452); the RAND corporation, 'to harness civilian science to military strategy' (Yarmolinsky, p. 58); the Advanced Research Project Agency, the Atomic Energy Commission . . .

Scientists learned that their career survival, the development of a research laboratory, could depend on how much it could be of interest to the military; and they quickly learned how to induce the military to *feel* that interest:

> An outstanding example of imaginative rivalry is the foundation of the Livermore Laboratory of the Atomic Energy Commission. While important work on nuclear warheads was done at the Commission's Los Alamos Laboratory, some scientists were dissatisfied with the speed of acceptance of new ideas there and organized the new laboratory in order to let their ideas have free play. Out of the ensuing rivalry came the speedy development of the hydrogen bomb and of many other devices which might have become available only after much longer time-intervals (Knorr and Morgenstein, p. 28).

An even more *outstanding example* of symbiotic relationship between scientists and military is the famous Jason Division of the

Institute for Defense Analyses.[1] IDA was used to address young scientists 'at a crossroads in their career':

> Consider IDA — an avenue worth exploring in your quest for professional advancement. IDA is an independent not-for-profit organization in Washington that performs significant scientific and technological studies on problems of national importance for the Office of the Secretary of Defense. . . . At IDA you're free from commercial pressure. You're free of vexing administrative duties that can cramp your effectiveness. Your whole intellectual capacity is free to focus on critical problems — giving them the full benefit of your technological expertise and analytical initiative... IDA can serve as a stepping stone in your career... Areas of interest where the value of your background and judgement is needed at IDA are: Tactical Systems, Strategic Systems, Sea Warfare, Weapons Effects, Advanced Sensors, Missile Defense, Space Technology... (IDA ad in *Scientific American*, November 1972).

IDA provided the frame in which the Jason Division was formed and employed by the military: 'the cream of the scholarly community in technical fields', 'a group of America's most distinguished scientists, men who had helped the Government produce many of its most advanced technical weapons systems since the Second World War, men who were not identified with the vocal academic criticism of the Administration's Vietnam policy' (*Pentagon Papers*, p. 120).

Early in 1966, a special scientific study group was assembled under the auspices of the Jason Division, to discuss 'technical possibilities in relation to our military operations in Vietnam'; the Jason report was given to McNamara in September. It advocated the construction of an electronic barrier along the demilitarized zone between North Vietnam and South Vietnam, a barrier in which fragmentation bombs and mines, including 'button bomblets' against foot traffic, played a major role.

The Jason Division has been asked many times to give advice and to provide help on issues of great relevance to the military; it is only one of the many bodies/agencies/research labs/think-tanks that channel and shape this fruitful symbiosis between scientists and military. The symbiosis is fruitful to scientists, as it provides easy funding of research, 'a stepping stone for one's career', prestige and personal power in the scientific community. It is fruitful to the military. On a more general level, it structures for them the institution of science. It also provides them with a body of scientists willing to listen to Reagan's latest demands: 'I call upon the scientific community who gave us nuclear weapons to turn their great talents to the cause of mankind and world peace: to give us the means of rendering these nuclear weapons impotent and obsolete' (*BAS*).

What Reagan is asking of his scientists is the development of beam weapons. The words about peace and the cause of mankind sound ironic in this context. And you can be sure that already his scientists are enthusiastically answering his appeal. Teller did not even wait for the appeal: 'The most important developments may come about in national defense. This probably will not mean bigger explosives. Defense against incoming missiles is more challenging, more important and more in accordance with what we wish to do' (Teller).

I think I have given enough examples to show that we can get rid of this idea of scientists as the critical consciousness of our society. That may be the image that they would like to give to others and to themselves, as when Dyson (a long-term member of the Jason Division) wrote: 'I believe that the work I have done as a Jason member has helped to strengthen the voice of sanity inside the American government' (Vitale, 1976); but the image cannot be taken seriously. If we want to understand the structure, the functioning, the peculiarities of the institution *science*, we have to think of the scientists as an organic part of this structure. The way they think, act, take initiatives, fight . . . all this is bound to the institution to which they belong and that has formed them. And this institution is more and more dependent on military funding of research. This objective dependence shapes the policies of the individual scientists and of their institutes.

What Do Scientists and Military Have in Common?

There is a non-negligible chance that mankind . . . faces the breakdown of the global order . . . The US superpower, cast by history in a role of world leadership, must be prepared to use its military force to prevent the total collapse of the world order (Parker).

Military research and development are necessary to acquire the technological strength needed to stabilize international security (Seitz and Nichols).

. . . in the first place, defence programmes provide security against external attack and help to maintain internal order and political stability, which are prerequisites for any economic progress at all (Kodzić).

I have a tentative, simplified, schematic answer to the question: what do they have in common, scientists and military? The answer is: the need for order and stability. G. Parker, of the first quotation above, is a political analyst; F.S. Seitz and R.W. Nichols are two very distinguished physicists; P. Kodzić is a social scientist at The Hague, where he was talking to fellow scientists during a disarmament symposium on 'The Dynamics of the Arms Race'. Quite different personalities, with quite different backgrounds; and then, they all seem obsessed with stability, security, world order . . . It is never explicit *which* stability (stability for South Africa?), *which* security (security for the government of Paraguay?), *which* world order (the order imposed by the two superpowers?) scientists are so eager to guarantee. But they are concerned with these concepts and — in this concern — there is something of the arrogance which makes many scientists convinced that they could deal with any problem in the world, as they have proved to themselves that they can deal with differential equations and genetic information.

Take as an example the long series of Pugwash Conferences devoted to 'Disarmament and World Security': Stowe, September 1961 (with the participation of Kissinger, who, however, 'did not join in the final resolution'); Dubrovnik, September 1963; Udaipur, January 1964; Sopot, September 1966; Nice, September 1968; Sochi, October 1969; Sinaia, August 1971 . . . Take as another example the series of Summer Schools organized by ISODARCO, the International School on Disarmament and Research on Conflict organized every other summer by 'the few physicists who constitute the core of the Italian Pugwash Group'. The themes are European security, terrorism, world security: Duinio 1970, on European security; Urbino 1974, on international terrorism and world security; Ariccia 1978, on energy and its security implications, and violence at sub-state level (*sic!*) . . . And the 'core of the Italian Pugwash Group' is not satisfied with their Summer Schools: 'ISODARCO is also running a research activity, on two small grants from the Italian National Research Council: the first on certain aspects of terrorist activity in Italy; the second on the physical security of fissile material.'

What are scientists so afraid of? Why do they need the shelter of security and world order so much? I think that a tentative answer has to be found in the ideology of the universality of science in its various changing forms: the international scientific community; knowledge being indifferent to national barriers . . . But this ideology has to be explained in turn. Scientists travel with more ease than most other people; they generally speak and understand a common language (at present, English); they are therefore led to feel that scientists in other countries are much nearer to them than, let us say, peasants and workers in their own country. World order, security and stability guarantee that the privileges and the power of some will not be challenged. And while scientists are not among the most privileged and the most powerful social groups in our societies (both East and West), they still have a lot to lose in any radical change.

It could seem that this last point, on the common interest of both military and scientists on order and security, is a minor one. It may well be so. But I have dwelt on it to give a concrete example of the way in which the ideologies and concepts essential for the survival of the ruling classes find their way among the social groups that share some of their privileges. The presence of a heavy component of military funding in research contributes to this transmission of values. This leads me to the last part of my talk: what is the research that is funded by the military? and why?

The Wide Spectrum of Military-Funded Research

> Modern technology and science are so complex and so interrelated that even in the final stages of the development of a weapon there is no necessary concentration on a specific 'military' technology (Knorr and Morgenstein, p. 31).

> It was recognized from the outset that the activities of the committee (the National Research Council) should not be confined to the promotion of research bearing directly on military problems, but that true preparedness would best result from the encouragement of every form of investigation, whether for military or industrial application, or for the advancement of knowledge without regard to its immediate practical bearing (Yarmolinsky, p. 286).

The military is funding research that is specifically designed to provide better weapons systems, i.e., those of larger mass destruction; research on parallel fields, whose results could lead to the discovery and development of new or more efficient weapons but are so fragmentary that they can be published openly with no danger; and finally research that is quite clearly and definitely of no military interest, be it direct or indirect. An example of the last category: a research on 'a model for pattern perception with musical applications', in which Bach's music is taken to pieces and analysed in depth (Rothenberg). What is the rationale behind this generous shower of funding by the military?

Among those who, under various perspectives, are engaged in a critique of science in our society and, in particular, in a critique of the role of military funding in research, there is a common misconception. Although you can never know how much military interest a potential area of research could have, it is untrue that 'you will never know' the hidden reasons that led the military, in some instances, to fund research that — to your naïve eyes — looks totally immaterial for military affairs (and I think that this last claim is too limiting). The quotations above make quite clear that the military is not only looking for explicitly military research and results; it is looking for a more general 'state of alert' of research and scientists which, in turn, generates 'sweet solutions' (as Oppenheimer once called them) and makes the military happy. The position of the NATO Science Committee is quite explicit about it:

> As the industrial democracies move from predominantly responding to the challenges and limitations of man's natural environment to fulfill the need for more effective control and management of the technological environment . . . science and technology have been necessary — if not sufficient — in the social transformations of the period since the Second World War. What science and technology have made possible is a continuation of the division of labour that is a distinctive feature of all society . . . We have some reasons to be confident that this tendency will continue. We have every reason to hope that it will . . . The vagaries of shifting national priorities have always affected alliances between nations, tending to make them short-lived and of decreasing value... When the North Atlantic Treaty was written, this had come to be well understood, and in extending the concept of mutual security to include co-operation in matters of social, economic, and political concern, it sought to widen the common interest of the alliance nations by strengthening and monitoring the stability of their institutions . . . In this way, the Scientific Committee would respond to the repeated request to shift the emphasis of its effort, to build bridges of co-operation between different scientific areas besides — as it did so effectively in the past — building bridges for the co-operation of scientists from different countries of the Alliance (Hemily and Ozdaś).

There is no doubt about it. Judging from the countries where some information is available (USA, NATO countries) the military is concerned with *all fields* of research; a healthy science (in the sense of an institution of science efficient, aggressive, collaborative) is a guarantee of power, control, potential mobilization. However, there is a point that still needs to be clarified: if the military is interested in the presence of a healthy scientific system in its country, why should they fund it? Why not leave the job to civil agencies? After all, they have the same aims: power, control, potential mobilization of the institution of science.

It is here that a specificity of military-funded research strikes our attention. It is true that all fields of research are potentially fundable by the military; but there are fields that are more so than others. In a subtle way, military funding alters priorities, emphasizes one field with respect to another and establishes 'military values' in the elaboration of a research policy. And, more than everything else, its value to the military lies in the web of dependences that it establishes among scientists. (It is significant that the work quoted from Hemily and Ozdaś, two senior NATO officers active in the Science Committee, bears the following two subtitles for the two volumes: 'Building on Scientific Achievement' and 'Technological Challenges for Social Change'.) It is this dependence that above all interests the military. When it becomes clear that the survival of a field of research or its development depends on the military, the corresponding scientists and technicians become willing to serve. The generosity of today is a good investment, in view of the domination of tomorrow.

What to Do?

It is difficult to arrive at any clear-cut answer to the question 'What to do?' Much depends on *who* is doing *what* and *why*. The community at large (for instance, in a small university town, around a large university campus) can hopefully struggle to demand that all research carried out at universities be non-secretive as to funding, employer, results and general policy frame. In particular, the presence in a campus of specific biological research oriented toward biological warfare could sound the alarm for many people to lead to a constructive and aggressive mobilization. I think that students and young research workers should *impose* their right to be explicitly and accurately informed about the research proposals to which they are subjected: only too often a young assistant is sent around to scientific Summer Schools or meetings on military funds about which he is completely unaware. I think *all* research workers, at any level, should refrain from asking for military funds. I know that the traditional alibi is 'If the military did not give me those $10,000, they would have bought another gun'; but it is a silly alibi. The military does not give out something for nothing; the accrued dependence of the Academy on the military is all in the interest of the latter.

To expose military-funded research in the universities could become an aggressive mobilization topic in several countries. However, I would advocate a line of argument and mobilization that avoids the traditional moralistic approach: the military is ugly, while civil research is for the common good of humanity. It is not true, and such an approach would give us a false start in the struggle. What is true is that dependence on the military could be harder to control or destroy than dependence on other civil agencies; that scientists involved in this kind of dependence tend to become more and more embedded in the military ideology; that the contribution of scientists is essential to the increase in power of the military; and many other reasons that, perhaps, could be elaborated starting from the few considerations above. But above all, there is a need for information about military-funded research in our research institutions. A detailed analysis of what is known — of what can be known, by searching patiently throughout the available material — can stimulate initiatives and new forms of struggle.

Notes

This essay has been published in a booklet, *Wissenschaft und Krieg* (Science and War), by the students union of the ETH-Zurich, Universitatsstr. 19, CH-8006 Zurich, from which copies are available for 10 Swiss francs. A version also appeared in *Wechselwirkund* 20 (February 1984), pp. 33–36. The essay is based upon talks given at the Polytechnic School, Zurich, 25 May 1983, and at the University of Strasbourg, 2 June 1983.

1.   Early in 1973, Jason left the Institute for Defense Analyses, which had been its home for almost 15 years, and moved into a new home created at the Stanford Research Institute in California. For more information. see Vitale, 1976.

References

A. Aigrain, in *Sciences* 79 (mars–avril 1971), quoting Neel.

Bulletin of Atomic Scientists, June/July 1983, quoting President Reagan's 'Appeal to the Nation' of 23 March 1983.

J.P. Baxter, *Scientists Against Time*, Boston, Little, Brown & Co., 1946.

P. Hemily and M. Ozdaś, *Science and Future Choice*, 2 volumes, Oxford, Clarendon Press, 1979.

H. Kellermann, *Der Krieg der Geister*, Weimar, Heimat und Welt, 1915.

K.E. Knorr and O. Morgenstein, *Science and Defense: Some Critical Thoughts on Military Research and Development*, Princeton, Center for International Studies, 1965.

P. Kodzić, 'Armaments and Development', in D. Carlton and C. Schaerf, eds., *The Dynamics of the Arms Race*, Croom Helm, 1975.

New Scientist, 22 September 1977, p. 738, 'Physicists try to forget Vietnam while promoting the neutron bomb'.

D. Noble, *America by Design: Science, Technology and the Rise of Corporate Capitalism*, NY, Knopf, 1977.

G. Parker, *Military Implications of a Possible World Order Crisis in the 1980s*, Report of the Rand Corporation, Santa Monica, 1977.

Pentagon Papers, Gravel edition, p. 120.

G.I. Powkrowsky, *Science and Technology in Contempory War*, Moscow, 1956, 1957 (in Russian); London, Stevens, 1959.

D. Rothenberg, 'A Model for Pattern Perception with Musical Applications', *Math. System Theory, 11,* 199 (1978); 'this research was supported in part by grants and contracts AF-AFOSR . . .'.

M. Schroder-Gudehus, *Les scientifiques et la paix: la communauté scientifique internationale au cours des anneés 20,* Montreal, Presse Universitaires, 1978, Appendix.

F.S. Seitz and R.W. Nichols, *Research and Development and the Prospects of International Security,* NY, National Strategy Information Center, 1973.

L.L. Strauss, *Men and Decisions,* Macmillan, 1962.

E. Teller, 'Role of Physicists in the 1980s', *Physics Today,* February 1981.

B. Vitale, *The War Physicists*; documents about the European protest against the physicists working for the American military through the Jason Division of the IDA, Naples, 1976.

B. Vitale, 'On the Neutron Bomb', *END Papers,* Winter 1981/82.

A. Yarmolinsky, *The Military Establishment,* NY, Harper and Row, 1971.

THE NEW WORLD AGRICULTURE GROUP: a History

Douglas H. Boucher and Isadore Nabi

Like most worthwhile undertakings, it began with vague feelings of discontent. We liked what we were doing, we did it well, and other people recognized it — all in all, we seemed to have a bright future ahead of us. And yet, somehow, it didn't seem right.

We were all leftist ecologists of one sort or another. We had been involved in lots of political activities at the University of Michigan and had come to feel a fairly strong sense of solidarity. We had stood out on picket lines in the rain for a month during the teaching assistants' strike in 1975; we had organized actions against sociobiology together; we had stood up for the rights of students, and against the various Mickey-Mouse acts of repression concocted by the Division of Biological Sciences. We weren't really organized but we were a group nonetheless. And as we worked together, the discontent slowly grew and became a force.

Its origin was the obvious and unconcealed contradiction between our beliefs and our daily lives. We knew and felt, based on long and hard political experience, that science was not neutral — that it was a basic part of the system which we were fighting, and which more and more of us were starting to call capitalism. And yet, we spent the greater part of our lives doing science — esoteric, academic, wild-things 'pure science'. Thus the feeling grew that perhaps if we believed all the political stuff we were constantly spouting, we should act on it. And not just in our 'spare time', or our Science for the People group, or after we got a permanent job, or any of the other accommodations we made with our consciences, but in what we revealingly called our 'work' — the research that took up so much of our time. We realized that, as students and professors, we had to do this to survive in the University — but maybe there was a way to do it differently.

In retrospect, we had already made a few halting attempts at it. Four of us had been involved in an Organization for Tropical Studies (OTS) field course in Costa Rica in 1974, during which we had done field problems on coffee picking and slaughterhouses, designed a research project on agricultural ecology, and made contact with the Costa Rican Socialist Party concerning helping them with some sort of technical aid. Many of us knew Dick Levins, who had been doing agricultural ecology in cooperation with Cuba since the mid-60s. And almost all of us were interested in the environmental movement and thus in ideas of organic agriculture. So, little by little our growing politicization led us to begin talking about changing our lives.

It began as weekly meetings for lunch up in John Vandermeer's office, without a really set agenda or purpose. My recollection is that it was Kay Dewey and Steve Risch who pushed it the most, in the winter and spring of 1976. Perhaps they felt the contradiction most, working as they had been on animal behaviour studies that were disturbingly close to sociobiology. But it wasn't just the behaviour people who had nagging doubts. I spent that summer trying to find ash and maple trees whose seed shadows I could measure, and repeatedly rewriting my thesis proposal. John Vandermeer, Steve Risch and others were in Costa Rica watching agoutis and tracking palm seeds, among other things. And, on and off, with varying casts and irregular frequency, the lunchtime meetings continued, until at some point (I can't remember any cathartic moment or blinding flash of revelation) we had decided, collectively, to do something.

The first concrete actions, however, were individual ones. Steve Risch scrapped a couple of seasons of bird-watching and came up with a thesis on insects in mixed cropping, and proceeded to Costa Rica to do it. He was soon joined by Mike Hansen. Their theses were the first of a still-growing series of studies of mixed cropping, a choice which remains a fundamental inspiration (or an obstacle to the further development of our thought, depending on your point of view). Kay Dewey abandoned a summer of watching marmots in the mountains of Washington, started taking courses in public health, and began searching for a new thesis topic. And as they showed that indeed one could change, it made it easier for the rest of us to act also. Indeed, it made it harder and harder for us not to change. For, beyond all the arguments that had crystallized our discontent and set our minds to imagining what ecology might be good for, we needed the argument of the deed — the persons who proved it was possible by doing it.

Agriculture was the obvious solution. Over the years we had come to think, correctly or not, that was a field fairly close to ecology, but *useful*. We had done some small-scale projects on it in Costa Rica as part of field courses, and we knew friends (Ron Carroll and Dick Levins, for example) who had applied their ecological ideas to agricultural situations already. Agricultural researchers used more or less the same theoretical framework as ecologists, it seemed, or at least they ought to. And it was clearly the basis of life of those Third World peasants who were both the most exploited and the most revolutionary class on earth. We already identified with them politically; agricultural research would allow us to do so in our daily life and work. Of course we would have to learn a lot of new information, but we could apply the same ecological theories and concepts.

Indeed, as we read more and more, it. became clear that agriculture not only could be studied with an ecologist's eye — it cried out for it. Those writers on agriculture with whom we felt the most sympathetic demonstrated very convincingly that agriculture under capitalism exploited not only humanity, but nature as well. The irrationality of U.S. agriculture — energy dependence, pollution, mechanization against workers, and all the rest — was merely an example of how capitalism regards the earth, as well as people, as a commodity to be bought, sold and used up. And so, the ecological consciences which we all had, as well as our social consciences, pushed us toward agriculture. We could finally integrate not only thought and action, commitment and work, but also society and nature.

The initial group effect came when John convinced the University of Michigan to give him a grant to try intercropping of annuals like corn and beans under a palm plantation in Costa Rica (at the La Selva Field Station in the rainforest, our old agouti-and-palm-tracking haunt). He proceeded to cut down several dozen disease-infested cacao trees on the Station to prepare the land for planting, and was told by OTS, which owned the station, that he hadn't gotten all the bureaucratic permissions necessary. Furthermore, after a great deal of effort was expended in digging trenches among the palms to plant a large experiment, most of our plants died. And so, although in the end John won his fight with OTS, we decided by the summer of 1977 to look for a new place to work.

The Road to Mexico

We were a 'we' by this time, although as yet unnamed. Most of us were involved in Science for the People, and so might have considered ourselves a subgroup of its Ann Arbor chapter, but we were still small enough not to feel the need to organize ourselves formally. But already we were attracting the interest of new people and starting to discuss projects with people we knew at other universities (especially SUNY at Stony Brook). We had also become more clearly focussed on the general question of mixed cropping (polyculture), which was predominant in Latin America, the apparent opposite of the conventional North American technology we had come to know and hate, and ready-made for an application of ecological ideas — diversity and stability, competition coefficients, the whole works. Several of us quite explicitly took the concepts and/or organisms we were interested in studying, and searched for some way to study them in Third World agriculture.

The search for a new site both made explicit the fact that we wanted to work collectively, and provoked the first extended discussions within the group about *how* to do research in agriculture (as opposed to *whether*). We had to decide between Michigan, the southern US (Florida and Texas seemed pretty tropical, which at that point was fairly important, for some reason), Mexico, and Central America, as well as between field stations like La Selva, agricultural experiment stations, agricultural universities, or some private eco-freaky farm. (As far as I can recall, we never considered choices that in retrospect would seem to have been logical, such as Cuba or peasant cooperatives.) In the end the choice narrowed down (if you can call it that) to agricultural stations and schools in Mexico — it was tropical, Third World, safe for leftists, and not too far away.

So, in the early summer of 1977 we wrote several dozen letters to all the Mexican agricultural stations and schools whose addresses we could find, asking if they would be interested in collaborating with a bunch of gringos. We got lots of replies, some puzzled, most positive, and planned out an itinerary that a group of us could take to visit them. In the end, there were four of us chosen to go on the 'Multiple Cropping Caravan', to visit all these places and see what we could set up.

As it happened, at just about the same time Steve Gliessman, whom many of us had known when he was managing an organic farm in Costa Rica, had been hired by the Departmento de Ecologia at the Colegio Superior de Agricultura Tropical (CSAT), in Tabasco on the Gulf coast of southern Mexico, where he had started a project to study the ecology of traditional mixed cropping. He had written a letter to the Institute of Ecology (INTERCOL) *Bulletin* inviting collaboration, and so we got in touch with him and arranged for the caravan to go to Tabasco as part of its survey of sites. The caravaners, after sending back a number of letters telling us how northern and central Mexico was a real desert, literally and figuratively, went to Tabasco and spent a couple of weeks looking at the CSAT, Steve's projects and the surrounding area. The CSAT is in the middle of a large, World Bank-financed development project called the Plan Chontalpa, and the *ejidos* (cooperative farms) of the Plan and nearby areas provided an interesting variety of social settings. Gliessman's experiments, involving multiple cropping and *chinampas* (raised fields), were along the same lines as we had been considering; furthermore, the climate was the sort of rainforest to which we felt accustomed from our years of working at La Selva. Also, several of the other faculty and students at the CSAT were leftists (many in the Partido Comunista Mexicano) who like us wanted to put their science to use in service of the people. All in all, it looked like just the right place.

Thus began 'the Mexico project', which persisted into 1981, through a variety of cast changes, renamings, reorganizations and internal divisions. When the caravaners returned we enthusiastically endorsed their recommendation to work with CSAT and began writing up grant proposals. While the first large one — involving John and Steve G. as Principal Investigators and several others as post-docs, students, or other such wage slaves — was not funded by the National Science Foundation, enough of us had other sources of money (fellowships, leftovers from other grants) that we could get started anyhow. Thus, more or less by accident, we began our tradition of diversifying our funding sources rather than depending on one enormous grant. Kay Dewey went back in the winter of 1978 to begin her thesis work, on child nutrition and cropping patterns in the Plan Chontalpa and nearby. The rest of us arrived by ones and twos over the summer of '78, mostly with the intent of spending a couple of months looking around, working on some experiments, and finding the Topic which we could both enjoy working on and justify to our universities as a Real Biological Thesis.

CSAT in Practice

From practically the beginning there were problems — the normal hassles of doing research in the tropics, of course, but also some new ones. The arguments were along several lines: socialist vs. ecologist orientations, Mexicans vs. gringos, social science vs. natural science, and pervading all of these, science vs. politics. The project wasn't like any of us had expected it to be, with some (in particular four of the Stony Brook students) being amazed that people wanted to stay up till midnight every night talking politics, when they had to get out in the field early in the morning, and others (especially Michigan folks) being amazed that such things weren't taken for granted. As the summer went on and the balance of people present shifted more and more to the second type, the arguments became more and more heated and personal. (Yes, the personal *is* political, but here the reverse became more and more true also.) Did everyone on the project have to have a Mexican collaborator? What did 'collaborator' mean? Should our Statement of Principles use words like socialism and imperialism? What did we need such a statement for anyhow? Over a variety of questions we played out the basic argument. In retrospect most of the splits could have been anticipated (though probably not avoided). Nevertheless, even leaving out the gossip, the misunderstandings and the pettiness, that summer was quite a shattering experience for a lot of us.

By the end of the summer, a couple of things had become clear. The group *had* to get together and define itself — not only to explain to Mexican leftists who we were and why we had arrived out of the northern blue (a dozen gringos, collaborating with an Ecology Department half that size), but also to prevent anyone from returning the next year without knowing what they would be getting into. To those of us who had begun the project for political reasons, and assumed that everyone else had joined it for them also, this self-definition was absolutely necessary, even if (indeed, partly because) it would exclude from participation some people who wanted to join us. On the other hand, this was (correctly) perceived as threatening the chances of apolitical people being able to do their thesis as part of the project, as they had expected — because others didn't like their politics, it seemed.

Our Mexican collaborators found this all pretty bizarre. There were some of the same reds vs. greens divisions among them also, but a far more important question was how they were going to benefit from our presence. Were we there to use the CSAT as a research site, or were we committed to sharing our knowledge with them? If we were, how were we willing to do this? And so began a second debate, which became even more important in 1979, about whether this was basically a research project or an effort to redress the intellectual exploitation of decades of American scientific imperialism. For some of us, it was quite a shock being called scientific imperialists because we wouldn't enter into teacher-student relationships with our Mexican collaborators.

Through all of this, we had still managed to find some pretty interesting things to study. In fact, one positive aspect of that summer (at least to those of us who have stayed in the group) was that we discovered that there are just as many exciting ecological questions in a corn field as in a tropical rainforest. True, most of us had chosen questions that were by and large pure ecology, with little social content. But that was seen as in most cases a necessary compromise with the requirements of our Biology Departments. Only later would new feelings of discontent with our apolitical, though agricultural, science begin to surface.

We left the CSAT at the end of summer 1978 with a Statement of Principles, some rules for participants in the project the following year, and a commitment to meet back in the States to revise these. We also had commitments for regular meetings of each 'chapter' at Michigan, Stony Brook and CSAT. We came up with a system of circulating curriculum vitae, research proposals and some sort of statement of political motivation among the chapters, to allow approval of people who wanted to go to Mexico the following summer, before they had actually packed their bags. And finally, we had a name: PACMISS, Proyecto Agricola CSAT-Michigan-Stony Brook-Stanford. The following spring we began our regular twice-yearly meetings in North America, which have continued every spring and fall in Ann Arbor, Boston, Montreal or Ithaca. These are generally mixtures of work (on various statements, rules, grants and papers), discussions (based quite often on position papers written in the chapters and circulated beforehand for discussion) and hedonistic reunions. We got started in a formal way with the Ithaca meeting in early 1979, which was a real high, for several reasons. The mere fact that we were meeting at Cornell (where Steve Risch had been hired a few months before), and that people had come from Boston, Mexico, and other places, showed that we were expanding beyond the original Michigan nucleus — that indeed there were other people who wanted to do this sort of thing too. Several of the people who came had backgrounds in fields like history of sociology, which added a new facet to our discussions as well as providing a new dimension along which to argue. Furthermore, a couple of Cornell people were actually in departments in a Faculty of Agriculture, even if they weren't really agronomists. An afternoon of seminars by various of us, which drew overflow audiences, demonstrated that Real Agricultural Scientists and Hot-Shot Ecologists might *both* be interested in what we had to say. All in all, it really made us feel like we had started something whose time had come.

On the other hand, some major problems came out concerning the following summer's research at CSAT. It became clear that (1) most of us hadn't gone through all of the approval process, and (2) no one was willing to tell someone else 'you can't go' despite their not having fulfilled the criteria. Furthermore, there was one person who came, an eco-reactionary of the pure right wing, who brought to a head the question of what to do about someone who says they accept a fairly progressive set of principles when to everyone else it's obvious that they don't. Attempts to be comradely and subtle succeeded only in having people decide to go to Mexico *and* expect not to get along there with those who thought they shouldn't go.

Summer of '79 in Tabasco had a new assortment of faces, but unfortunately some of the same old hassles. The 'how-often-do-we-have-to-talk-politics?' and 'what-if-I-can't-find-a-collaborator?' questions resurfaced, and the gringo-Mexican split came to a head over whether we were willing to set up a system to teach them what we knew. And a further problem came up — how and when do we start actually working with peasants? We had a substantial set of experiments set up on the Colegio, and several people were doing social science projects, interviewing people around the Plan Chontalpa, but this didn't seem to be quite what 'science for the people' meant. The peasants were the *object* of our research rather than subjects. Finally, for some of us (i.e., at least for me) the 'traditional agriculture', going back to the ancient Mayas, which we were supposedly studying, seemed more and more like wishful thinking. The picture of Tabasco agriculture which the social science people in our group were developing had a lot more to do with modern capitalism than with ancient traditions.

On the other hand, we did get a lot of good research done that summer, at least by traditional scientific criteria. One of the points of origin of the project, Kay Dewey's thesis, turned out to be a brilliant example of how one could integrate biological and social concerns and still get a Ph.D. and a job. The personal interactions within the group (most of us shared the same house) were generally pretty cooperative. And most important, on July 19th of that year the Sandinista Revolution was victorious in Nicaragua.

Bringing It Home

It took us quite a while to make the formal contacts with the new Nicaraguan government concerning the possibility of working with them, even though several of us had collected money for the FSLN in Michigan before the triumph. We were talking more and more about the difficulties of doing research which serves the people in a system which doesn't, but in the meantime we continued on as the 'Mexico project', through a variety of name changes — to PACITH, Proyecto Agricola Cooperativa Inter-disciplinaria en el Tropico Humedo; then GIANM/NWAP, Grupo de Intercambio Agricola 'Nuevo Mundo'/New World Agriculture Project, and finally NWAG, New World Agriculture Group (pronounced 'newag') at the start of 1980.

That school year saw further expansion (I came to Montreal and started a little group, and people from California and North Carolina got in touch with us, while when Ron Carroll and Carol Hoffman moved from Stony Brook to Texas we essentially lost one group and started another). It also saw the crystallization of many of our arguments in the series of papers originating from Michigan on 'transitional technology', which developed into 'technology and class'. These provoked reactions from various of us, especially those in social science and/or Ithaca, who felt that the argument betrayed an appalling ignorance of political economy. At the same time, we realized more and more that we needed some new ideas to develop our practice further, particularly since we were now beginning to work in North American agriculture as well as in the Third World. The support work for the Farm Labor Organizing Committee (FLOC) in the Midwest, to which several of the Michigan group had become very committed, clearly involved important technical questions as part of the politics — the mechanization of tomato harvesting above all. It was clear how technology was being used against workers, but could we, as workers in technology, help develop a technology that favored the working class?

As that debate went on, another idea also began to grow — to start trying actively to spread our message. Thus far we had mostly spoken to each other, and to committed friends, but the increasing interest was a sign that what we were trying to do struck a responsive chord with a lot of people. And so, in the fall of 1979 we started talking about organizing a major conference on ecology and agriculture at which a larger group could discuss the issues we were dealing with. The exact purpose of the conference, and therefore its size and format, was discussed at length. The two poles of opinion were: a major international event at which well-known people from both ecology and social science would give papers, versus a well-prepared NWAG meeting at which we would discuss our analysis of world agriculture in depth and try to clarify our group's position. In the end, after many discussions, the decision was more or less made for us by our inability to get grants to fund a big conference. The meeting, called 'Alternatives in Agriculture' (ironically, we had chosen a non-political title to avoid scaring off potential granting agencies), was finally held in Montreal in May of 1981, and at the least got us back to talking about politics. For in the intervening months we had put so much effort into 'work' activities — chiefly, planning the conference and the summer's research — that discussions had taken a back seat.

By summer of 1980, the project had spread out from Tabasco — several people worked at the INIREB station in Jalapa, Veracruz (a much more comfortable climate) instead of CSAT, and people at Michigan began an intercropping experiment with tomatoes and cucumbers. Steve Risch at Cornell and I at McGill also set up intercropping experiments, but the Michigan ones were somewhat different in that they were (1) practically unfunded, and thus (2) dependent on the work of a great many undergraduates, and most importantly (3) justified not only by the supposedly more ecological and progressive nature of intercropping, and the possibility of getting useful comparative information for the Tabasco work, but also by the link to FLOC's struggle. By showing that intercropping of tomatoes and cucumbers, which *had* to be harvested by hand (we thought), was more profitable for farmers than the potentially mechanizable monocultures of tomatoes generally produced in Ohio, NWAG could help FLOC members preserve their jobs by slowing down the process of mechanization. Or, if the multinationals like Campbell's and Libby's continued to favor monocultures despite our scientific evidence against them, this would be a powerful argument that they wanted to mechanize not to increase productivity, but rather to prevent farm workers from organizing. Thus, the mixed cropping research was an example of the idea of transitional technology.

Needless to say, this interpretation was controversial. But the Michigan experiments in 1980 were important as an attempt to make our science serve people in North America as well as Latin

America, and the transitional technology position was clearly a step forward in explicitly analysing agriculture in terms of class. Ironically, it was becoming clear at the same time that our ideology was outrunning our practice. As in the early days, we were talking about a new relation between research and politics, but continuing to do the same old things. Those were, basically, mixed cropping, biological control, and socioeconomic studies *of* (not *with* or even *for*) farmers and peasants. We were accumulating publications, and later grants, and in general succeeding in our field by traditional scientific criteria. But the work on the side of the oppressed remained far away from our research.

In different ways, the same tensions were developing in Tabasco. We gave some mini-courses at CSAT and carried out a series of experiments on one of the *ejidos* near the Colegio, but it had become clear we were really giving very little to our Mexican collaborators, and practically nothing to Mexican peasants. Furthermore, despite our nice scientific results showing the value of mixed cropping, the growing of corn and beans in any way, let alone together, was getting harder and harder to find in the Plan Chontalpa. And we knew why, since the social scientists among us had shown clearly how the cropping patterns of the heavily-indebted *ejidos* of the Plan were determined by the desires of the banking system and associated government agencies, rather than by ecological rationality. Our political analysis made it obvious to us why our research efforts were futile.

These conflicts were brought to a head by Steve Gliessman's decision to leave the CSAT for the University of California at Santa Cruz at the end of 1980. This made it clear that it was he who had been our major 'Mexican collaborator' in the project. Only two of us returned to the CSAT in the summer of 1981. Instead, we began to think about other places to work, and, in line with our discussions culminating in the Alternatives in Agriculture meeting, about which classes to ally with and support. In the North American context, this created a difficult and as yet unresolved dilemma: where, in a class analysis based on Marxism, did the 'family farm' fit in? Were family farmers basically exploiters (as it seemed in Ohio, where the farmworkers we were supporting were in direct conflict with them) or exploited? Was the 'family labor farm' a transitional form, destined to disappear as farming was industrialized, or a permanent part of capitalist agriculture? And what did this imply about supporting organic farming, which often substituted hand labor for chemicals? These questions were discussed much, though not always well. They are still with us.

The Nicaragua Connection

In the meantime, we had finally gotten in touch with the Nicaraguans. Through our Costa Rican contacts with the FSLN, and through my talking with the Nicaraguan consul in Montreal in November 1980, we set up a visit to Nicaragua by a half-dozen of us in early 1981. The group visited several experiment stations and talked to friends who now were serving in the Nicaraguan government, and came back filled with enthusiasm. The Reagan administration's action against Nicaragua and El Salvador helped push us along; at our meeting in Montreal in spring '81, five of us were chosen as an official NWAG delegation to go to Nicaragua and work out a collaboration in August 1981.

Nicaragua was a revelation. To those of us who had worked in Mexico or Costa Rica, the differences were incredible. Here was a country where the government was favoring the workers and peasants, and where the old exploitation of nature for profit was something not only to be denounced, but to be changed. Indeed, by a completely different path than we had taken, many Sandinistas had developed ecological ideas about agriculture, and they were trying out ideas — mixed cropping, biological control, research with peasants — which we had been trying to encourage in Mexico without the slightest success. Furthermore, they needed and wanted our help; both the economy and the development of science lagged well behind Mexico and Costa Rica. It seemed a place where we could finally put our beliefs into practice.

At the same time, we had to be cautious, if only to avoid repeating the mistakes of the Mexican collaboration. There, the result of our failures had simply been personal conflicts and a failure to sustain the tie. But we knew that if we screwed up in Nicaragua, it would be more serious, particularly for our Nica comrades. Furthermore, it would be more important than before to be sure about who we sent down, since a group like ours would be a useful cover for US government infiltration. As Galio Gurdian, a friend from Michigan who now directed the Nicaraguan Center for Studies of the Atlantic Coast (CIDCA), explained it by analogy with a disease of tobacco recently discovered in Cuba, 'We don't want you to send us the blue mold.' So we set up a set of steps that had to be followed in order to work in Nicaragua under NWAG auspices. Bureaucratic-appearing as they were, and although we really had no control over people just going to Nicaragua on their own if they didn't like them, such steps seemed the minimum necessary to build a solid collaboration.

Our first NWAG collaborator in Nicaragua was Charlie Hale, who stayed in Nicaragua after the delegation left and began to work with Galio at CIDCA. The success of his work, and the trust we accumulated by our support work from North America (sending down literature, applying for grants, speaking and writing about the Nicaraguan Revolution) made it possible to send down several others in the summer of 1982. When Charlie left for Stanford in August '82, Galio requested that we find another NWAGgie to replace him, and Kathy Yih was picked for the job. I returned in July–August 1982 and set up a long-term collaboration with the Soil Fertility Department of the Ministry of Agriculture, as well as with the Investigaciones Regionales office.

Inevitably, some of the questions that we had dealt with in NWAG before — ecology vs. production, the meaning of collaboration, the proper role of gringos in Latin America — have arisen again in connection with Nicaragua. And there are others that are specific to a small nation in revolution and under attack. Nonetheless I think we can say, at the least, that we have learned something from our previous experiences. We will no doubt make mistakes, but at least they won't be the same ones.

As the Nicaragua collaboration has developed and grown, from one initial cooperant to nearly a dozen, it has tended to influence the structure of the organization more and more. Indeed, our collaboration has forced us to adopt more structure, simply in order to be able to deal effectively with the Nicaragúan government and its bureaucracy. Thus, we have set up first a Nicaragua Coordinating Committee, and then a NWAG Steering Committee; we have a Coordinator in Managua, and a small office there, and are preparing official reports on our activities. At the same time, we have retained a decentralized structure in that important decisions are still made by our whole-group meetings, generally by consensus. We still run on extremely limited funding, and each local chapter has its own activities and operates fairly independently (and sometimes out of communication) from the rest.

While our work in Nicaragua has been tremendously exciting, and has gained us a good reputation there, there has been a growing dissatisfaction during 1984 with its diffuse nature. Cooperants go down to work on individual, separate projects with different institutions; though many of us have lived in the same house, our working days are quite separate. And while many of us have contributed substantial resources (books, copies of articles, equipment, even pens and pencils) to the institutions with which we work, there has been little general mobilization of NWAG as a whole to support Nicaragua materially as well as intellectually.

This has finally, at the end of 1984, given rise to a series of decisions which are likely to transform our work in Nicaraguan substantially in the near future. We are undertaking to start a major support project in Nicaragua *as a group*; to raise substantial donations for it in money and in kind; and to use it as a vehicle for involving many more people in our support work. This is partly a response to Nicaragua's increasing needs, and partly to our own; as we have found ourselves being contacted by more and more persons who have scientific training and want to help Nicaragua, we have found it difficult in many cases to suggest something they can do to help.

NWAG's Future

Ironically, despite our feeling that we have not been doing as much as we could in Nicaragua, we have also realized that the Nicaragua work is taking up more of our time and effort. In particular, the analysis of North American agriculture has been put off, by and large, since the Alternatives in Agriculture conference in 1981. While the polyculture, biological control, and similar sorts of ecological agriculture projects have continued, perhaps because of academic inertia as much as anything else, the political discussion of why we do this work has lapsed. Thus, one of our tasks for the forthcoming year is to reanimate this debate, with (what we hope will be) a more developed perspective than before.

NWAG is clearly in a growth phase today, with new chapters in Vermont, Minnesota, Washington and elsewhere and a regional structure (East, Midwest and West) with alternating regional and international meetings every six months. While certain groups have gone through extended periods of being out of contact with the others, each of the twice-yearly meetings has seemed to have involved new faces from new places. We are finally becoming a continent-wide organization. While we have made some adjustments, such as setting up a newsletter to replace our irregularly sent and usually indecipherable minutes of group meetings, we've retained the loose framework of the early days. The meetings produce somewhat less heat than before; arguments are less gladiatorial, sexism is less blatant, at least, and we domineering-professor types have learned to restrain our tongues and our body language. On the other hand, the weekends are as packed with activities as ever, and one of their major functions is still that of solidarity — reassuring each other that we have, and are, comrades.

Notes

This history was written by Doug Boucher in April of 1983 in response to a request from the rest of the group; it has been modified and brought up to date for publication with the help of comments and corrections by several other group members. I thank all those who contributed, particularly those who did so in writing; they helped correct many errors in chronology and interpretation. Nevertheless the responsibility for this version is mine alone.

The notes of the 1989 Annual Conference of NWAG were submitted by Isadore Nabi in response for a request by the Montreal chapter, which hosted the spring 1984 meeting, that each chapter brainstorm about what they would like NWAG to be, five years in the future. This was intended as a way of organizing a discussion of the group's political orientation and future direction. Unfortunately no other chapter responded in writing before the meeting, so the planned discussion took a different form; however Nabi's notes give a good sense of what a NWAG international meeting is like.

NWAG polyculture experiments at the University of Michigan Botanic Gardens, 1981

THE 1989 ANNUAL CONFERENCE OF NWAG

As recorded by Notetaker Isadore Nabi

The 1989 meeting was held in April at Lafayette, Indiana, as proposed by the Purdue Chapter, and then shifted to Ames, Iowa, because that is what some people thought we decided last time, but those people left before the end of the 1988 Conference so they don't really know.

Friday pm	Open symposium jointly sponsored by the Entomology Department and NWAG: 'Polyculture and You', followed by a Potluck on the shores of the Wabash. Music was provided by the NWAG Winds, and ended with the (disputed) NWAG anthem, 'Meet Me at Eclosion Time in Chalatenango'.
Sat. 8.30 am	Bagel and cream cheese breakfast (Bagels brought by Chicago Chapter).
9.30–12.30	First Plenary 1. Go around the room introducing ourselves 2. Criticism and self-criticism 3. Discuss when to take a break 4. Take break
12.30	Break for lunch. Actually it was 1.15. Reconvene at 2.00 (actually 3.30).
3.30–5.30	Meeting of Standing Committee (floating workshops) 1. Small farms or collectives? 2. The past of NWAG 3. The present of NWAG
6.30	Supper (Cassava and rice, prepared by the Cassava/Rice Taskforce). Some members to see 'The Seeds of the Seedy', the first full-length documentary based on the 1988 NWAG Conference. Some won't. The Wild Party, the Polyculture discussion and the simultaneous Plenary Sessions on Chapter Reports and Who is a Member were held on the busses.
9.00 pm–6.30 am	Busses to Ames, for rest of meeting. Party and Plenaries and Polyculture discussion group. Proposal: Why NWAG should organize in the pharmaceutical industry.
Sunday 6.30 am	(Actually 1.30 am because of flat at Galena, but really 9.30 because we crossed the time zone): Bagel breakfast at the Waxy Kernel, eatery of the Genetic Engineering Department.
8.00–10.30 am	(Actually 11.00–3.00). Plenary session. Criticism, self-criticism and recrimination. Expulsion of Purdue Chapter. Slide Show: NWAG Tour Through Liberated Zone

of Haiti.

3.00–5.00 pm   Plenary
1. Newsletter
2. Polyculture
3. Principles of Unity
4. Statement of Principles
5. Debate on incompatibility of Principles of Unity and Statement of Principles
6. Report on the Purina Boycott
7. Pizza
8. Newsletter reminder and threat
9. Proposal of Saginaw Chapter on revision of logo
10. Criticism (we passed the point of self-criticism an hour ago).

5.00 pm
1. Motion for adjournment from Gainesville Chapter
2. Motion ignored
3. Chapter reports
4. Polyculture
5. Who is a member of NWAG?
6. Political Principles
7. Newsletter
8. Gainesville leaves
9. Criticism of Gainesville
10. Motion to Adjourn from Regina Chapter
11. Motion amended to add word 'later'
12. Reports from Standing Committees
13. Polyculture
14. 'Are we radicals, leftists, or socialists?'
15. Polyculture
16. Notetaker leaves for regional meeting at Lafayette

ONE-DIMENSIONAL PLANNING THEORY

Don Parson

the plan has been made
ideas emotion logic and substance
have all been collected in a model
— apologies to the Minutemen

The claim of planners and planning theory to scientific expertise
as a justification for plans made and actions taken is a hallmark of
the profession. It is a connection that is used to damn planners
and planning theory — and rightly so — from liberal, marxist and
anarchist perspectives (e.g. Davies, Castells, Kravitz, Goodman).
With epistemological forays into its scientific origins, the role of
science in planning — indeed, of planning *as* science —is
seemingly accepted (see Camhis, Los, Boddington).

   In this paper I will try to move beyond 'use-abuse' models of
planning as science by suggesting that our conception of science
itself severely limits our capacity for liberatory self-activism. This
limitation is not due to the *misapplication* of either science or
rationality by planners. Rather, the scientific worldview itself —
what William Blake described as 'single vision' — postulates a
single, objective reality that is to be appropriated through the
scientific method. By accepting planning as science, 'alternative
plans' (especially those of the left) seek only a reorganization, and
not a transcendence of, or opposition to, the current reality. In
contrast, I will argue for a multi-dimensional planning theory that
is not based on science as the sole means of understanding and
constructing reality.

Planning and Science

Comprehensive social planning via Keynesianism and the
welfare state first emerged with the growth of working-class power
in the 1930s. Prior to this, planning was confined to the factory.
There the planning process was strictly defined by the scientific

method, with Taylorism being the foremost example. Where it did venture outside of the factory walls, as in the City Beautiful efforts at the turn of the century, planning was embroiled in an academic debate about 'was it art?' or 'was it science ?'. 'Efficiency-minded municipal engineers struggled against commercial planners who were more interested in beautifying the city's facade than easing its vehicular circulation' (Boyer, p. 67).

The scientific basis of planning was firmly established when, during the Depression, Karl Mannheim defined planning as 'rational decision-making'. This definition was the foundation for all subsequent bourgeois planning theory, which has sought to build a rational consensus for a technocratic model of societal development (see Parson). In the seminal *Retracking America*, John Friedman defined planning as the 'process by which a scientific and technical knowledge is joined to organized *action*' in order to impact society (p. 246).

The critique of bourgeois planning by the left, with proposals for the development of socialist planning, has been historically two-fold. Prior to the 1930s, the left had assumed that planning was the definition of socialism (e.g., Engels). A planned economy, where goods were produced and allocated on the basis of human needs, was antithetical to the 'anarchy' of the capitalist mode of production. Choices were clear — planning or capitalism; rationality or anarchy.

The capitulation of capital to the working-class struggles of the 1930s in the form of Keynesianism forced the left to recognize that capital could, and did, plan. Anarchy was not the definition of capitalism any more than planning was the definition of socialism (Panzieri). Writing in 1948, C.L.R. James recognized the extent to which the working class had forced capital to plan, and that planning was not the essence of socialism: 'Only the proletariat can free the plan; otherwise the plan reproduces with intensified murderousness all the evils of value production' (p. 155).

Thus 'contrary to popular mythology, planning did not bring socialism — in fact it became a sophisticated weapon to maintain the existing control under a mask of *rationality, efficiency and science*' (Goodman, pp. 171–72, my emphasis). What is important to note here is not the abandonment of the planning vs. capitalism or rationality vs. anarchy dichotomies, but their replacement with that of a socialist rationality vs. a capitalist (pseudo?) rationality. It now becomes possible to construct a simple use-abuse model of planning: 'To truly disentangle planning from its conservative tendencies, it must exist within a political economy based upon socialist principles of distributive justice' (Beauregard, 1978, p. 250). Capitalism abuses and perverts planning while socialism would take advantage of its proper use. The slaughter of millions of peasants that took place when the GOSPLAN was implemented in the USSR can be reduced to a technical problematic or an error in procedure because ultimately, as Blazcya informs us, 'planning [under socialism] is good for you'.

The above calls not for a further delineation of rationality or a refinement of the scientific method in the service of socialism, but for the questioning of science as the foundation of planning theory. The scientific worldview postulates the existence of a single, objective reality that exists as separate and distinct from the subjective activity of the inhabitants of that reality. This is a radical alteration of the worldview of preceding epochs where 'reality' was composed of myriad subjective' cosmologies, often based on organic metaphors (Roszak, Merchant).

Indeed, one of the central problems in the emergence of science was the eradication and destruction of alternative cosmologies in favor of objective reality. Feuer notes that the initial social acceptance of science in England lay in Elizabethan and Reformation coffeehouses. Coffee was praised by the emerging capitalists as a 'wakeful and civil drink', not 'unfit for business' as were the traditional ale, beer or wine (p. 56). The alertness and wakefulness induced by the consumption of coffee allowed one to bypass or ignore 'lesser' cosmologies in favor of the 'real world'. As one member of the Royal Society claimed, 'coffeehouses had improved useful knowledge, as much as they [the universities] have . . .' (p. 54). Perhaps this seems a trite example, but it is being used to illustrate a point: nearly 300 years later when J. Robert Oppenheimer witnessed his handiwork with the explosion of the first atomic bomb, he recalled the line from the Bhagavad-Gita, 'I am become death — the shatterer of worlds' (Feuer, p. 394). Is Oppenheimer's observation an illustration of a 'good' methodology gone awry? Or it is an extension of the 'single vision', the nascent form of which was seen in English coffeehouses of the 17th century?

The scientific method sees rationality as the mechanism by which to appropriate objective reality. Here the connection between science and planning — as rational decision making — is made explicit. Rationality certainly pre-dates science, but in the preface to his *Principia* Newton saw rationality (as well as mechanics) as integral components of the 'new philosophy'. According to Greek philosophy, one could comprehend the truth through rationality only if one were a moral person. An immoral person, no matter how rational, could not understand truth. Descartes, and other advocates of the new philosophy, advocated the rational method of discovering the truth which could exist independently of whoever applied that method. Whether moral or immoral, anyone could appropriate the truth by means of a scientific rationality (Michel Foucault in Rabinow, pp. 371–72). Science cannot exist in the absence of rationality, but not necessarily *vice versa*. To 'act rationally' and to 'rationalize something' become nearly indistinguishable. To paraphrase Feyerabend, rationality becomes a game used by intellectuals to befuddle the natives — one makes plans 'for' people while convincing them to act appropriately because that is the 'rational' thing to do. Again, this is not an aberration, but the essence of the scientific method.

With a planning theory based on science, the use-abuse model is the starting point for alternative plans. But by accepting the equivalence of planning = rational decision making = an appropriation of a single, objective reality through the scientific method, alternative plans can only utilize a 'better' science in order to facilitate a more rational (or better rationalized), more efficient (and, by implication, better managed), organization of current reality. And that current reality — the single vision, the one and only objective world — is, of course, our existence as moments in the capital relation.

A brief example might help clarify the above assertion. In Communist-controlled Bologna transportation is free, work is being shifted from profit-oriented tasks to socially useful labor, and the town is planned on a human scale — all in a manner far more rational than would be possible with the unbridled anarchy of capitalism (Jaggi et al.). But the bottom line is that one must still go to work, only now you can get there 'free' (and on time) thanks to the planned traffic system; one still lives in a housing project, but now a humane one, etc., etc. In this context, socialist planning appears as the reorganization of work-oriented society — a more rational and more scientific form of existing reality.

The concept of *interest*, that is, the rationalization of a diverse constellation of the needs, wants, passions and desires of those for whom plans are being made, is the means by which rational decision-making can be operationalized. Bourgeois planning theory has postulated the existence of a 'public interest' as a technical problematic which might reduce this diversity to a plannable variable. Bourgeois critiques might question how to determine the public interest, but not the concept itself.

Similarly, planning theory from the left has transformed the concept of public interest into that of 'class interest'. Substituting the use-abuse model of planning in lieu of class analysis, planning for the public interest is seen as suffering from an imprecise definitional delineation such that special interests (capitalistic ones) are inevitably, though perhaps unintentionally, served. With the proper determination of the 'correct' interest of the working class, however, planning can serve the interest of the class by building socialism.

I hope that my criticism of planning theory is clear: I am not questioning the abuse of the criteria of rationality or the abuse of the way that criteria are operationalized via the concept of interest. Rather I am questioning whether a planning theory that is based on science, that is, the assumption that there is a single, objective reality to be appropriated by rational means, is tenable in light of current working class movements. As I will suggest later, the recent cycle of struggles indicates that there is not one reality but many different worlds that are being actively created, implying that there is no single class interest.

Based on the foundation of science, left planning can only reorganize of rationalize existing reality. Thus it is an extension of bourgeois planning theory, representing what Herbert Marcuse described as 'the closing of the universe of discourse' (p. 84). Planning theory becomes one-dimensional, as can be seen by examining 'alternatives'.

Programmatic Alternatives

The uni-dimensionality of planning theory can be illustrated by considering alternative planning proposals. I want to briefly examine those proposals being made to deal with the current crisis of capital: reindustrialization, democratic planning, and the appropriation of planning technology (e.g., information systems) by workers to be turned against capital.

Reindustrialization is a plan that is being called for across the political spectrum. Those involved in the 'reindustrialization debate' (Wolff) are in agreement that the growth of diverse movements in the 1960s and early 1970s, characterized by a specialized constituency (e.g., women, blacks, Chicanos, students, gays, etc.), an unwillingness to participate in the labor–liberal coalition, and the utilization of dissensus politics, has caused a profitability crisis for capital through the blockage of Keynesian development (see Parson). As a result, we need 'to establish a modicum of speedy, disinterested decision-making capacity' (Thurow, p. 16).

Reindustrialization strategies on the left mimic those of Thurow but in a more 'humane' manner, seeking to channel the energy and autonomy of popular movements into electoral politics with the goal of economic democracy (Carnoy and Shearer, 1980). The capital-labor relationship is accepted and it is argued that the problem of capital will be solved with an extension of wages, benefits, low-interest home loans and the welfare state, as contented workers make productive workers (Bluestone and Harrison). Participation in existing reality, in the capital-labor relation, is rationalized by the reasoning of 'we are in a crisis' and therefore 'we have no option' (Hearn, p. 168). Planning is the mediating factor between capital and labor, and central to the process of reinstituting a societal consensus.

Consensus appears to be the common denominator of all planning proposals. The Trilateral Commission calls for 'democratic planning' that will be 'carried on in such a way that people will be involved in helping to set goals. This is a preferred alternative to some kind of technocratic elite model' (Crozier et al., pp. 175, 179). Left alternatives seek a 'democratic-consensus model' of planning which proposes to 'transform the entire basis of legitimation, democratizing the control of capital, thereby replacing private accumulation with democracy as the decisive constraint of policy' (Kraushaar and Gardels, p. 292).

The imposition of consensus over capital illustrates the uni-dimensionality of planning theory and its left alternatives. Capital is the single, objective reality — the real world of the scientific method. The public or class interest is 'served' by instituting a democratic consensus as a rational policy alternative to the exploitative excesses of capital. But this is a consensus over capital and thus the democratization and rationalization of our own exploitation. As Marx pointed out, '. . . the idea held by some socialists that we need capital but not capitalists is altogether wrong. . . . The concept of capital contains the capitalist' (p. 512).

Information technology is becoming increasingly central to the planning process. It would seem that, with a planning theory founded on science, the blueprints for the New Jerusalem might appear on a computer print-out. The idea that we can appropriate this technology for 'our side' in the class struggle assumes not only that technology is neutral, but it is also statist in that 'strategy is dependent on the planning and management capacities of the (externally situated) state' in order to administer that technology (Robins, p. 10). And, as Robins has shown, information technology has and is being used to integrate both centralized and decentralized decision-making into a systematized 'information grid'. The planned application of information technology is thus central to the re-establishment of capital's control over the social factory.

The above dovetails with democratic planning and reindustrialization alternatives. Decentralized decision-making founds a new phase of capital accumulation, administered by the state in either the public or class (take your choice) interest, with a democratic control over the entire process. The end result is more capital and more state and — no matter how rational — more of the same. Block has described this 'paradox' of alternative planning: 'When the left trims its sails to develop programs that fit the orthodox model, it becomes as practical and realistic as Reaganomics' (p. 74).

Towards a Multi-Dimensional Planning Theory

I will suggest here the possibility of a planning theory neither based on science and its worldview nor rationalized by the concept of interest. I do not question the existence of planning, but its expression in the form of rational decision-making. Planning is undertaken by both capital and the working class — the former to realize value by subjugating the totality of human experience to the one-dimensional existence of work; the latter to assert its own self-defined needs, wants, passions and desires (Parson, Phillips). Not to recognize the class basis of planning is to retreat either to use-abuse models of planning or to the planning/anarchy dichotomy, but with anarchy arising from the unprogrammable self-activism of the working class rather than from the capitalist market. In other words, I am arguing for anarchy *and* planning and not anarchy *versus* planning. This is possible with a planning theory not founded on science.

The history of the development of the scientific method highlights whether the scientific worldview is compatible with any sort of non-hierarchical and non-exploitative social arrangement. As Carolyn Merchant has shown, while science successfully challenged certain hierarchies in the medieval social order, e.g., the power of the clergy or the divine right of monarchs, it did so by creating and enforcing other hierarchies. This can clearly be seen in Francis Bacon's technocratic utopia of 'New Atlantis' where the 'new philosophy' of science was inapplicable to irrational and passive entities such as women or nature. As a court attorney, Bacon was an avid prosecutor of witches — throwing some light on the social motivation of an avowedly neutral and objective methodology.

The erosion of medieval hierarchies by the scientific method did not unleash the possibility of liberatory development, as there was a desperate rush to maintain order by replacing the obsolescent organic worldview with a newer mechanical metaphor. Indeed, science has assumed (even required) a 'natural' order to exist. It is questionable whether such a methodology can comprehend the 'irrational', the spontaneous, or forms of anarchy or autonomy. As Bacon stated in his explanation of the scientific method: 'the mind itself [must] be from the very outset not left to take its own course, but be guided at every step, and the business be done as if by machinery' (cited in Roszak, p. 150).

Trying not to resort to marxology, it seems to me that, by way of a highly subjective reading of the *Grundrisse*, one of the methodologies that Marx was developing was not an analysis of capitalism as it exists in the objective world. Rather, for Marx objectivity appears as a reflection of the reality that capital has succeeded in imposing upon us. The destruction of that reality is a consequence of our asserting (many) alternative realities through the self-definition of our own needs, wants, passions and desires.

The creation of such cosmologies can be thought of as a very succinct description of working class planning. Workers are not only those that receive a wage, but include all those whose labor has been annexed by capital in the 'social factory' (Cleaver, 1979, Dalla Costa and James, Baldi). Thus the working class plans to be, paradoxically it would seem, non-workers, that is, to be no longer a moment in the capital relation. By contrast, capitalist planning revolves around the imposition of the one-dimensional reality of work. In this context, working class planning is the development of prefigurative politics — ways of living that transcend domination, authority, hierarchy, capital and the state. In other words, it is the refusal of the 'single vision' — the objective reality that capital, through science, has imposed on us.

As such, the autonomy, spontaneity and diversity of the recent (1960s to early 1970s) cycle of working class struggles (plans) demonstrated an explosion of experimentation with alternative ways of appropriating, creating and changing reality. Blacks,

Chicanos, women, students, lesbians, the disabled, youth, wild-catting factory workers, the counter-culture — all rejected a partnership with capital and the state to define themselves not as workers but according to their own positive criteria *as* blacks, *as* women, *as* part of the counter-culture, etc. There was the creation of myriad alternative and oppositional cosmologies, none of them strictly subordinate to the one-dimensional reality of capital. (I am reminded, from 'Star Trek', of the Vulcan definition of beauty — 'Endless Diversity in Infinite Combinations'.)

This seriously calls into question the concept of interest. If there is a single working-class interest, it is expressed negatively as the opposition to capital. (This can currently be seen in the blockage of Keynesian development from which capital has not yet recovered.) Beyond this there are a multiplicity of self-defined needs, wants, passions and desires to be liberated. Working class planning is thus characterized by the liberation of diversity and not the imposition of unity. This cannot be accounted for by alternative plans like those already discussed here, which seek to make us all democratic workers, instead of realizing the liberatory and prefigurative potential of being a non-worker.

How is a planning theory to be neither founded on the scientific worldview nor constrained by an interest to be planned for? Central to such a project is the comprehension that science and its attendant rationality are neither the sole mechanism nor the final word on means to appropriate, create or alter reality. Poetry, art and ritual predate science in this matter, and I hope that these means, among others to be developed, will postdate science as we actively create our own cosmologies. 'As each man and woman becomes a poet and a visionary, they also become open to the true scientific spirit. People spontaneously create science just as they spontaneously create art' (Kovel, p. 212).

As such, science (like art) *is* social relations: 'we must learn to theorise our own practice and to practice our own theories — to combine the vision of an alternative world with prefigurative struggles in our own daily lives . . .' (Young, pp. 113–14). My argument, then, is not simply anti-science. Rather, science should be one means (among many) in a non-hierarchical array of methodologies to realize our self-defined needs, wants, passions and desires. Neither is it humanist in the sense of leaving a sterile and lifeless universe of motion and matter to the realm of science while other methodologies might be considered more appropriate to human social systems. We must change the scientific method and its one-dimensional reality just as we create our own cosmologies. In such a manner we might 'utilize technology and science as moments of our own self-activity in struggle against capital and for the creation of our own kinds of worlds... — spaces that prefigure and create the post-capitalist world' Cleaver, 1983, p. 98).

Capital generates a powerful and oppressive reality that cannot be wished away by retreating into our own worlds — we must assert, sometimes violently, our realities. Though we continually struggle against it, work is done for capital to get money, to buy food, clothes, etc. (We are all familiar with the routine of selling our lives in order to buy back commodities.) With this in mind, one might ask how a multi-dimensional planning theory would apply to 'actually existing planners' — those professionals working in a planning office or doing the work of teaching planning. A multi-dimensional planning theory might aid in the questioning of science in planning and amongst planners:

> By conceding the epistemologically objective status of science, radical professionals set limits on how far they challenge their own complicity in reproducing forms of power which scientific and other professional practices carry and constitute... It is all too easy for people to privilege the practice of professionals, as workers with a particular (if contradictory and unclear) class location, without analysing the political conditions in which the ideas themselves are produced (Radical Science Journal Collective, 1981, p. 45).

While I have maintained that there are plans of the working class (in the sense of prefigurative politics), there are no planners 'for' the working class. This is a task performed by ourselves, individually and collectively, as we realize our own diverse cosmologies. There can be no working class planning departments, implying the need for a continued dehierarchicalization and deprofessionalization of planning (McDougall). In free societies, Feyerabend speculates:

> Problems are solved not by specialists (though their advice will not be disregarded) but by the people concerned, in accordance with the ideas *they* value and by procedures *they* regard as most appropriate... They gradually conquer the free space that has so far been occupied by specialists and they try to expand it further. Free societies will arise from such activities, not from ambitious theoretical schemes (pp. 9–10).

This is a long way from seeing planning as the application of science to social problems by means of rational decision-making.

Conclusions

I have argued that a planning theory founded on science is one-dimensional because it assumes the existence of a single, objective reality to be appropriated by rational means. By contrast, the prefigurative struggles of the working class indicate the construction of many diverse realities. Rationality is a dubious criterion from which to assess these struggles. Planning theories that are founded on science extend or rationalize the objective reality imposed by capital. A multi-dimensional planning theory would seek to accommodate the development of myriad alternative cosmologies — a 'radical pluralism' — asserted through the self-defined needs, wants, passions and desires of the (ex-) working class.

References

G. Baldi, 'Theses on Mass Worker and Social Capital', *Radical America*, *6*, (1972) 3–23.

R. Beauregard, 'Planning in an Advanced Capitalist State', in R. Burchell and G. Sternlieb, eds., *Planning Theory in the 1980s*, New Brunswick, NJ, Center for Urban Policy Research, 1978.

G. Blazyca, *Planning is Good for You: The Case for Popular Control*, Pluto Press, 1983.

F. Block, 'The Myth of Reindustrialization', *Socialist Review*, *14* 1 (1984), 59–76.

B. Bluestone and B. Harrison, *The Deindustrialization of America*, New York, Basic Books, 1982.

S. Boddington, *Science and Social Action*, Allison and Busby, 1978

M.C. Boyer, *Dreaming the Rational City*, Cambridge, MA, MIT Press, 1983.

M. Camhis, *Planning Theory and Philosophy*, Tavistock, 1979.

M. Carnoy and D. Shearer, *Economic Democracy: The Challenge of the 1980s*, New York, M.E. Sharpe, 1980.

M. Castells, *City, Class, and Power*, Macmillan, 1978.

H. Cleaver, *Reading Capital Politically*, Austin, University of Texas Press, 1979.

H. Cleaver, 'Thoughts on RSJ 11 Editorial', *Radical Science Journal* 13 (1983) 97–100.

M. Crozier, et al., *The Crisis of Democracy: Report on the Governability of Democracies to the Trilateral Commission*, New York University Press, 1975.

M. Dalla Costa and S. James, *The Power of Women and the Subversion of the Community*, Bristol, Falling Wall Press, 1972.

J. Davies, *The Evangelical Bureaucrat*, Tavistock, 1973

F. Engels, *Socialism: Utopian and Scientific*, Peking, Foreign Languages Press, 1925.

L. Feuer, *The Scientific Intellectual*, New York, Basic Books, 1963.

P. Feyerabend, *Science and a Free Society*, New Left Books, 1978.

J. Friedmann, 1981 *Retracking America*, Emmaus, PA., Rodale Press, 1981.

R. Goodman, *After the Planners*, New York, Touchstone, 1971.

J. Hearn, 'Decrementalism: The Practice of Cuts and the Theory of Planning', in P.

Healy et al., eds., *Planning Theory: Prospects for the 1980s*, New York, Pergamon Press, 1982.

M. Jaggi et al., *Red Bologna*, Writers and Readers, 1977.

C.L.R. James, *Notes on Dialectics*, Westport, Conn., Lawrence Hill and Co., 1980.

J. Kovel, *Against the State of Nuclear Terror*, Pan Books, 1983; Boston, South End Press, 1984.

R. Kraushaar and N. Gardels, 'Towards an Understanding of Crisis and Transition: Planning in an Era of Limits', in C. Paris, ed., *Critical Readings in Planning Theory*, New York, Pergamon Press, 1982.

A. Kravitz, 'Mandarinism: Planning as a Handmaiden to Conservative Politics', in T. Beyle and G. Lathrop, eds., *Planning and Politics: Uneasy Partnership*, New York, Odyssey Press, 1970.

M. Los, 'Some Reflections on Epistemology, Design, and Planning Theory', in M. Dear and A. Scott, eds., *Urbanization and Urban Planning in Capitalist Society*, New York, Methuen, 1981.

G. McDougall, 'Theory and Practice: A Critique of the Political Economy Approach to Planning', in P. Healy et al., eds., *Planning Theory: Prospects for the 1980s*, New York, Pergamon Press, 1982.

K. Mannheim, *Freedom, Power and Democratic Planning*, New York, Oxford University Press, 1950.

H. Marcuse, *One-Dimensional Man*, Boston, Beacon Hill, 1964.

K. Marx, *The Grundrisse*, Penguin, 1973.

C. Merchant, *The Death of Nature: Women, Ecology and the Scientific Revolution*, San Francisco, Harper and Row, 1980.

I. Newton, *Mathematical Principles*, Berkeley, University of California Press, vol. I, 1962.

R. Panzieri, 'Surplus Value and Planning: Notes on the Reading of *Capital*', in *The Labour Process and Class Strategies*, CSE Books, 1976, pp. 4–25.

D. Parson, 'Plan and Counterplan: Notes on Moving Beyond the Crisis in Planning Theory', in *Contemporary Crises* 9 (1985), 55–74.

R.J. Phillips, 'Global Austerity: The Evolution of the International Monetary System and World Capitalist Development, 1945–1978', unpublished doctoral dissertation, University of Texas at Austin, 1980.

P. Rabinow, ed., *The Foucault Reader*, NY, Pantheon, 1984.

Radical Science Journal Collective, 'Science, Technology, Medicine and the Socialist Movement', *Radical Science Journal* 11 (1981), 3–70.

K. Robins, 'Capital and Cable', Cable Working Papers 2, prepared for the GLC and Sheffield City Council, 1983.

T. Roszak, *Where the Wasteland Ends*, Garden City, NY, Anchor Books, 1973.

L. Thurow, *The Zero-Sum Society*, Penguin Books, 1981.

G. Wolff, 'Reindustrialization: A Debate Among Capitalists', *Radical America*, 15, 5 (1981), 7–15.

B. Young, 'Science *is* Social Relations', *Radical Science Journal* 5 (1977), 65–118.

RADICAL SCIENCE AND THE MODERNIST DILEMMA

David Dickson

A few months ago, *The Economist* carried a provocative cover article entitled 'How Europe has failed'. The drift of its argument was that a worrying gap was growing between the European economies and those of the United States and Japan, and that this gap was a direct result of the Old World's reluctance to grasp and exploit the opportunities offered by modern science-based technologies with the same degree of single-minded enthusiasm as its two economic competitors. However, all was not lost for Europe. A few outposts of full-blooded enthusiasm for the miracles offered by high technology could still be found. And the magazine ended its six-page article by quoting the words of Carlo de Benedetti, chairman of Olivetti, that in these enclaves of future hope the new technologies 'are taking the drags off the wheel of human activity and letting it move as fast as human imagination can spin it'.

If such sentiments are familiar to anyone who has been following the recent debates about the central role of high technology in advanced capitalist societies, the language in which they are expressed should be equally familiar to historians of art. It also poses vital questions for the radical science movement. For it is almost identical to the wide-eyed enthusiasm for technology preached with religious fervour by a variety of art movements in the past, and in particular by the Italian futurists, who flourished briefly but provocatively in the years immediately prior to the First World War. Marinetti, one of the leaders of the movement, wrote of the 'new beauty' of the machine, suggesting that 'a roaring motor car, which runs like a machine gun, is more beautiful than the *Winged Victory of Samothrace.*' An early Futurist manifesto proclaimed that 'universal dynamism must be rendered as dynamic sensation; that movement and light destroy the substance of objects'. Indeed, the Futurists went as far as to inform their followers that 'the triumphant progress of science makes changes in humanity inevitable, changes that are hacking an abyss between those docile slaves of tradition and us free moderns who are confident in the radiant splendour of our future'.

It has been tempting to dismiss the overblown rhetoric of such statements — indeed, the whole philosophy of the Futurists — as little more than a misguided infatuation with machinery and the world of modern science. There is certainly much to criticize in the subsequent evolution of the more widespread cultural and political movements for which they helped to provide the groundwork. One was the whole Bauhaus tradition of modern design, which argued that social artefacts should be designed to meet narrowly defined criteria of social function, and which injected what has been called a 'machine aesthetic' into all the familiar objects of everyday life. More pernicious was the role played by Futurism in the rise of Italian fascism. Marinetti, for example, who had written that socialism was wrong because it was 'international and unpatriotic', and had suggested that 'Italy shall be governed by a government of twenty experts, vitalized by an assembly of young men', became an early supporter of Mussolini and, as a result, was later appointed secretary of the Fascist Writers' Union.

However there is, I suggest, a need for the radical science movement to examine the ideas and ideals of groups such as the Futurists more deeply than a superficial assessment might indicate. For, as suggested by the quotation above from Benedetti of Olivetti, the rhetoric once associated primarily with marginalized groups of cultural radicals has now become one of the chief ideological planks of advanced capitalism. An uncritical enthusiasm for high technology has been a key factor used by the Reagan administration in the United States to generate a broad consensus behind a wide range of its recent policies. At one extreme, stringent limitations have been placed on the effectiveness of regulations protecting the health and safety of workers on the grounds that such regulations hold back the rate of technological innovation. At the other, the same message about the inherent desirability of a high-tech future has been used to defend plans for the giant space station which are being proposed by the National Aeronautics and Space Administration. This project is not needed by anyone in the US except those companies who expect to win lucrative government contracts to build it, and the government itself, which hopes that a bold new space initiative can distract attention from the social disruption caused by its earth-bound programmes.

European politicians have not been slow to follow Reagan's example. In Britain, a Conservative government which has presented strong ideological reasons for allowing market forces to dictate the decline of traditional industries, such as coal and steel, has had no such reluctance in justifying its intervention in many areas of the new 'high technology'. In some cases, such as the Department of Trade and Industry's Alvey programme in advanced microelectronics, it has defended its actions on the grounds of an eventual commercial pay-off. In other cases, of which participation with other European governments in the construction of the US space station is perhaps the most obvious example, this excuse cannot be used (since neither the industrial nor the scientific 'community' feels it is a worthwhile project), and government officials have to admit that the whole scheme is being pushed for political and propaganda reasons.

Nor is the strategy limited to right-wing governments. When French President Francois Mitterrand reshuffled his government in June 1984, the key theme that he demanded his new ministers should pursue was the 'modernization' of French society. Chosen to spearhead this campaign was the 38-year-old technocrat Laurent Fabius, previously Minister of Industry and Research, who quickly and eagerly took up the President's challenge. In his first major speech to the National Assembly outlining the policy directions that he intended his new government to pursue — a speech which one commentator later described as a 'hymn to modernization' — Fabius made it clear that Mitterand's appeal had two objectives. The more pragmatic was an attempt to generate a political consensus behind the government's efforts to restructure its industrial base, moving away from traditional industries into the brave new world of microelectronics and biotechnology. (Revealingly, Mitterrand's announcement some weeks earlier that the government had decided to close down a substantial part of the French steel industry was made only a few days after he had visited Silicon Valley in California.) The second goal, which Fabius also acknowledged, was an attempt to regain some of the socialist government's rapidly-failing popularity among the middle classes by arguing that modernization was a non-political and exciting objective in which all classes shared a common stake, and whose pursuit was therefore in the equal interests of all.

Or take recent proposals that Umberto Agnelli, Vice-President of Fiat in Italy and one of Europe's foremost capitalists, made to a meeting organized last year by the Council of Europe:

> It seems likely that over the next few decades we shall see a world situation in which a few major cultural areas will confront and influence each other. In this scenario, Western civilisation will certainly play a leading role, exerting considerable influence on other cultural models. Inevitably, however, this will lead to a clash with a variety of different impulses from Third World countries, the Japanese sphere and so on. In this clash of cultural diversities, science, especially as applied to technology, will play a unifying role to an extent unparalleled in the history of mankind. New technologies will truly be the carriers of a common language.

Contemporary examples like these demonstrate how the message of the Futurists has been given a new lease of life in what diplomats call 'the highest political circles' as part of capitalism's cultural response to its challengers in the mid-1980s. The central message being preached by capitalism is not that technology is neutral, but that technology in general — and high technology in particular — is exciting and desirable. If we want to move forward in our understanding of the political role played by science in modern society, then it is crucial to understand this shift and its implications, if only for the reason that so much of the analysis of the radical science movement over the past decade and a half has directed its critique at the ideology of science as neutral. Focusing our critique on the supposed neutrality of science and technology, I suggest, risks a serious misdirection of effort. What we should be looking at, in contrast, is the resonance between the images which society creates of scientific/technological activities and the meaning given to these images in the consciousness of the individual; the way this resonance appears to make the activities themselves desirable and thus builds support for the political framework that underpins them; and finally, the type of cultural/ political intervention that would be required to break this circle and to demonstrate the potential superiority of different technological projects based on different political assumptions.

As one step in this direction, for example, we must try to understand how the very success of the Futurists in stimulating a critical assessment of prevailing cultural assumptions — or, in reverse, of the Reagan administration in convincing the voting population of the 'commonsense' validity of its high-tech rhetoric — raise important questions about the political use of images of technology. Do we accept that the rhetoric of high technology inevitably reinforces those political forces which are currently the most successful at exploiting it? Or can we build a politically progressive case around the potential benefits of the new technologies? That is, given that establishment forces herald them as ushering in a new era of human prosperity, should we argue that this prosperity will be achieved only if certain political barriers are removed? (It should be remembered that several leading members of the Futurist movement identified themselves as social, as much as artistic, revolutionaries.)

Origins of Modernism

Some guidance can be found by situating the debate about the cultural role of modern science and technology within the broader context of current controversies over 'modernist' movements more generally. The purpose of this brief essay is to suggest some of the ways in which these apparently separate debates link up. In particular, I want to suggest that both the successes and the limitations experienced by the radical science movement of the 1970s were similar to those of the so-called 'post-modern' movement in other fields of cultural practice, ranging from painting to architecture. By engaging in and learning from the debate that has recently been taking place between supporters of modernism and post-modernism at the cultural level, we can gain some useful insights into the political struggles over the economic and political modernization of Western capitalist economies.

There is a sense, or course, in which almost all ideas identified with science for the past three hundred years might be characterized as 'modern'. The Scientific Revolution of the seventeenth century carried within it the germs of the idea that the wisdom of the classical philosophers — the ancients — was not a rigid boundary to people's thoughts, but could be improved on by modern man through the power of the scientific method. In the eighteenth century, the philosophers of the French Enlightenment turned this conviction into a political programme. Writers such as Condorcet and Diderot argued that the rapid development of science and its technical applications would inevitably lead both to material and — equally importantly — to moral progress. The philosophy of Francis Bacon and the scientific theories of Isaac Newton both appeared to reinforce the claim that science represented the epitome of modern thought.

Such was the intellectual power awarded to the natural sciences that they became accepted as the epistemological basis of the Enlightenment's own brand of philosophy — positivism. Indeed, the link between an uncritical confidence in the powers of science, and support for 'positivistic' theories of social organization and social interaction, remains a powerful one today. We find it, for example, in any attempt to organize collective activity — ranging from factory production to the delivery of health care — on 'scientific' grounds. And the ideological nature of such arguments has been one of the key targets of the post-1968 radical science movement. Indeed, frequently included in these attacks on positivistic thinking have been the ideas and strategies of earlier left-wing movements among scientists, which often rested on the argument that the rational or 'scientific' organization of society was a prerequisite for the proper exploitation of science and its application to social goals.

Yet modernism — at least in the way the term is conventionally used to describe a cultural style — cannot be crudely identified with the scientistic spirit that infests positivism. For the modernistic spirit was born out of the cultural struggles that took place in the middle of the nineteenth century, and in particular out of an effort to define the arts as a sphere of consciousness that stood *in opposition to* the social impact of industrialization and rapid technological change, at the same time as acknowledging the extent to which it was inevitably embedded in these changes (in contrast to the romantic movement, which tried to distance itself from them). Its target was those — including virtually the whole of the positivist school — who confused material and spiritual progress. Its central belief was the idea that art was needed to address the growing imbalance between the two. One of the first 'modernists' in this sense, for example, was the French poet Baudelaire who, in a typical passage, suggests:

> There is . . . a very fashionable error which I am anxious to avoid like the very devil. I refer to the idea of 'progress'. This obscure beacon, invention of present-day philosophising licensed without guarantee of God or Nature — this modern lantern throws a stream of chaos on all objects of knowledge; liberty melts away, punishment disappears . . . Take any good Frenchman who reads his newspaper in his cafe, and ask him what he understands by progress, and he will answer that it is steam, electricity, and gaslight, miracles unknown to the Romans, whose discovery bears full witness to our superiority over the ancients. Such is the darkness that has gathered in that unhappy brain.

The distinction that the early modernist artists and poets such as Baudelaire drew between the material world and the spiritual and aesthetic world can therefore be seen at one level as a reaction to efforts by scientists from the seventeenth century onwards to divide sensory experience into primary and secondary qualities.

The first of these referred to the objective characteristics of the material world which were equally accessible to all — such as mass and force; the second referred to the subjective experiences stimulated differently in each individual by these characteristics, a metaphysical split (described in detail by A.N. Whitehead) whose long-term implications remain central to the preoccupations of the radical science movement of today.

The central dilemma facing the modernist movement arises from the fact that it has sought not to deny this split but to exploit it. Indeed, so influential have its efforts been on the subsequent development of European culture for the past one hundred years that the identification of 'pure art' as a category of cultural practice has remained central to what Raymond Williams describes as the new 'structure of feeling' that came into existence in the middle of the nineteenth century as a by-product of the Industrial Revolution.

Three separate strands can be distinguished that connect the nineteenth-century debates about the relationship between art, science and technology to our concerns of today. The first is the extent to which, as suggested above, those who espoused the modernist movement — at least initially — did so as critics of the existing social and political order. It was the archaic political structure of the time, they frequently argued, that was holding back the full exploitation of science in the broadest interests of human beings by overstressing the purely material aspects of technological progress. This was the line of reasoning, for example, picked up most strongly by the Futurists (and, a decade later, by those Russian constructivists who saw in their art a reflection of the promise that science held out to the new socialist state).

The second strand in the argument comes from the fact that just as central to the modernist revolution as the concept of 'pure art' was the concept of 'pure science', the idea that it was possible to describe science as a purely intellectual activity with no reference to previous moral, religious or philosophical traditions. It is significant, for example, that the word 'scientist' was coined at precisely this time by the Cambridge mathematician William Whewell. According to the Oxford Dictionary, this occurred only eight years after the date on which the word 'artist' was first used to distinguish someone who practised the 'fine' arts, as opposed to the 'industrial' arts. In other words, it seems likely that the same social roots lay behind the identification of the arts (by 'modernists' such as Baudelaire) as an autonomous sphere of cultural practice and the identification of the sciences no longer as a branch of natural philosophy but as an independent sphere of intellectual practice characterized, above all, by their claims to objective knowledge.

Finally, it seems important to understand how this new 'structure of feeling', in which the 'pure subjectivity' of the arts and the 'pure objectivity' of the sciences played such a central role, was itself not merely a reflection of the various values respected by the new industrial capitalism of the mid-nineteenth century, but a central element of the strategy by which this capitalism redefined — and thus reproduced — its social and political relationships. From this point of view, the famous 'two cultures' split between the sciences and the arts which has been discussed so much in the past few years appears not as an unfortunate by-product of the process of industrialization, but as an essential component of the political context within which this process took place.

Much of the history of Western culture — or, at least, of the dominant ideas about Western culture — over the past century reflects the interaction between these three factors: the assumption of the autonomy of the arts as the realm of the aesthetic/subjective; the assumption of the autonomy of the sciences and their technological applications as the realm of the material/objective; and the use of the two-dimensional cultural space created by these complementary assumptions to describe the 'apolitical' field in which other forms of cultural practice, ranging from music to architecture, are supposed to take place. Indeed, in many ways it is the interrelationships between these three factors, rather than any one of them on their own, that has determined the overall structure of 'modernist thought'.

The evolution of this set of modernist assumptions is conventionally traced through its impact on the arts alone. One dictionary of modern thought, for example, describes modernism as 'the comprehensive term for an international tendency, arising in the poetry, fiction, drama, music, painting, architecture and other arts in the West in the last years of the 19th century and thus affecting the character of most 20th century art'. As a result, the history of modernism tends to be understood primarily as the history of a lengthy series of art movements, starting from symbolism, impressionism, fauvism, cubism, post-impressionism, futurism, constructivism, imagism and vorticism in the period leading up to the First World War, and leading on to expressionism, dada and surrealism in the post-war period.

Similarly, the so-called 'post-modernist' label which is often attached to cultural movements of the past decade is also confined in its application to the arts, where it has been used to identify those movements explicitly seeking to reject the assumptions of the modernists. We find it applied most strongly, for example, in painting and in architecture. In the first of these it tends to be used to describe a return to pre-modernist forms of naturalism, the idea that the value of a painting lies not merely in the aesthetic values expressed by the image on the canvas, but the broader social, cultural and historical values that this image invokes in the consciousness of the spectator. The same is true in architecture. Here, too, we find the label 'post-modernist' being applied to the work of those who explicitly reject the formalist styles of their modernist predecessors, seeking to inject what is often described as a more human dimension into the design of buildings. This is achieved again by explicitly linking the design into its social, cultural and historical setting, reflecting and reinforcing this setting rather than attempting, through architecture, to create something which both reaches beyond and reflects back on it.

If we consider science and technology to be culturally neutral, then they have no place in the history of cultural movements and styles apart from their use as tools and techniques for expressing ideas and values that exist at a deeper, more subjective level. If, on the other hand, we accept that science and technology are, in themselves, important carriers of the modernist spirit, then we must include them as integral parts of the history of modernism in their own right. Of course, we can only do this if we are prepared to admit that the *idea* of the objectivity of science is itself a cultural artefact; this means, in fact, that we must reject any claims to absolute value made by the whole modernist project, since part of this is itself based on the concept. However, once we have accepted that the strategies of this project have been largely determined by industrial capitalism — whether they are the strategies of capitalism's supporters or opponents — then we are free to argue not merely that public projections of science play a political role as a component of national ideology, but that this role can be properly understood only by placing it in the wider context of that played by modernist ideology in its broadest sense.

Two more recent events seem to support this line of reasoning. The first has been the coincidence between the post-modernist trends which emerged in the arts in the 1970s and movements in scientific and technological circles epitomized by labels such as 'social responsibility in science' or 'alternative technology'. The central strategy of these latter movements was directly to challenge the modernist presentation of science and technology, i.e. the idea that the value of both could be legitimately assessed independently of their social context. The political movements around science and technology that originated in this period have sought explicitly to locate both activities in their social, cultural and historical settings. On the one hand it has been argued that science and technology inevitably reflect the values of their environment (the arguments about 'science and technology as ideology'); on the other, that as social practices they should be infused with a more explicit awareness of their links to this environment (reflected, for example, in demands for more 'science, technology and society' courses for undergraduate scientists and engineers in universities and polytechnics). Taking a cue from the way in which very similar arguments were taken on board by the arts, it does not seem unreasonable to describe this as the 'post-modern movement' in science and technology.

The second factor is the reverse side of this coin. There have recently been several critical analyses of post-modernism, in the conventional cultural sense, arguing that the post-modernist critique of modernism is superficial, and that as a result its political impact tends to be reactionary. Post-modernism is attacked for concentrating on the surface phenomena of cultural experience, rather than the deeper ideas and values which these phenomena express. The links which it claims to make to the surrounding environment serve not to make the political role of cultural artefacts more transparent but, in contrast, to mask it even more completely. In other words, it divorces the debate about cultural modernization from that of social modernization. Where the modernists argued that the two formed complementary components of a single movement, the post-modernists in contrast have frequently argued that it is possible to pursue the first (in its new post-modernist mode) while ignoring the second.

Jürgen Habermas, for example, explores this argument in detail in a key lecture entitled 'Modernity vs. Post-modernity'. He claims that 'rather than giving up modernity and its project as a lost cause, we should learn from the mistakes of those extravagant programmes which have tried to negate modernity'. The main problem, suggests Habermas, is that the project of modernity has not yet been properly fulfilled:

> The project aims at a differentiated relinking of modern culture with an everyday praxis that still depends on vital heritages, but would be impoverished through mere traditionalism. This new connection, however, can only be established under the condition that societal modernisation will be steered in a different direction.

Again there is a significant congruence that we can draw upon between these criticisms of post-modernism as a cultural style and critiques of the post-modernist trend in science and technology. Whatever the original political goals of the alternative technology movement or that which pushed for more social responsibility in science, their thrust has been blunted by the co-option of their more superficial arguments. We see this process working, for example, in the way the concept of alternative technology has been transformed into the less challenging idea of appropriate technology, and as such has been taken up by multinational corporations and institutions which express their interests, such as the Organization for Economic Cooperation and Development Similarly, demands that the scientific community should be more responsive to the social dimensions of science and technology have been absorbed into a variety of 'technology assessment' or 'technology regulation' institutions devoid of any political challenge to dominant forms of decision-making.

Radical Science Debates

It is in this broad context that I want to turn now to the two books that have served as a stimulus for this essay, Chris Jones' *Essays in Design* and Marshall Berman's *All that is Solid Melts into Air*. Their relevance to the argument outlined above is that, taken together, they illuminate from opposite directions what appears to be a critical boundary in our present thinking about radical science. The issue that both books raise is the extent to which we should remain within the post-modernist paradigm that has determined much of our strategy over the past fifteen years, or in contrast, should revitalize the modernist paradigm which we originally rejected (by dismissing, for example, the ideas of the previous generation of radical scientists, such as J.D. Bernal). Jones' book places the goals and aspirations of the alternative technology movement, of which he was conceptually one of the early theoreticians, directly and explicitly into the framework of post-modern culture. Berman, in contrast, defends the spirit of modernism — including that which, he claims, is captured by the promise of science — against the post-modernists. Like Habermas, he argues that the problem is not modernism itself but the way that it has become absorbed into the ideology of capitalism. Neither book provides radical science with an easy path through the modernist dilemma; but both, I suggest, provide us with clues to some of the directions that we might take.

At one level, these books are very different from one another. *Essays in Design* is a collection of articles, newspaper interviews and miscellaneous writings by Christopher Jones, an industrial designer who has long been raising questions about the inappropriateness of 'scientific' approaches to industrial design, indeed, of rigid social structures in general. (Jones resigned in the mid-1970s as the first professor of design at the Open University, where he had been responsible for several innovative technology courses, but after he eventually found 'that that too had become rigid and inhuman'.) Marshall Berman is professor of political science at the City University of New York; his book *All that is Solid Melts into Air* — the title is a phrase taken from Marx, who uses it to describe the constant changes in the material conditions of modern society — is a lengthy analysis of the idea of modernity in Western political and literary thought, concentrating on the influence of key individuals (ranging from Goethe through Marx and Dostoevsky to the American highway planner Robert Moses), and arguing from the activities of such figures that radical critics of the idea of modernity have been misguided in their target. The real struggle, according to Berman, is not to reject modernity as the window dressing of advanced capitalist society — even if this is what it has become — but to rescue it for socialism from those who have diverted it to capitalist ends.

Despite the difference in focus, there are also some immediate similarities and common themes between the two books. One is the way in which they both take the city — the dominant man-made object of our time — as one of their main points of departure. The city has frequently been analysed as merely a product of the capitalist economy, the urban centre which, on the one hand, houses those who work for capital in its factories and offices, and on the other exercises a hegemonic control over the surrounding countryside, which it gradually moulds to meet its own demands (e.g. as a resource for the production of food) and its own values (e.g. as a safety valve offering temporary relief from the pressures of urban life).

In contrast, both Jones and Berman come not to bury the city, but to praise it — at least to praise what it could become. For them, the city is the melting pot of all human relations, the privileged site not only of class conflict but of class solidarity and interpersonal relations, the environment in which all important human experiences take on their most intense form.

Ironically, it is Jones, the professional designer, who takes the more overtly philosophical approach. In the mid-1960s, Jones wrote a widely quoted paper for an architectural journal in which he invited speculation on the desirable forms that cities should take in the future. In sharp contrast to the dominant ideas among professional planners of the time, Jones argued that the design of a city should start from a genuine assessment of the needs and desires of the people who were likely to live in it, not form an abstract idea of what, in the planners' minds, a city should look like. The article helped inspire a generation of architectural students to search for new, 'bottom up' approaches to planning problems, indeed to the whole question of how technology can be used to met social needs more effectively. (One indirect offshoot, for example, was the alternative technology movement, which started life in the basement of the Architectural Association in 1969.)

Jones' essays trace the development of the various themes which brought him to this position and beyond — indeed, which have determined his attitude to design in general ever since he learnt the limitations of functionalism at first hand in the late 1940s. One is the conviction that designing is — or should be — not an aesthetic exercise but 'a way of improving relations between objects and people'. Another theme, which carried a familiar ring for those engaged in the debates of the radical science movement over the past decade, is his concern that 'industrial life, considered as a set of organized procedures, excludes at almost every point . . . the freedom to act as a person and not as a thing'.

But Jones has also remained sensitive to the other side of the coin: namely, that even when such criticisms are accommodated by the design profession (and its paymasters) — an important factor in stimulating the style that has become known as post-modernism — the new design methods supposed to meet the previous deficiencies are often as bad as the ones that they replace.

> My intention was to find ways to make the design process more sensitive to life, but what happened was the imposition of methods that were larger in scale than those which we had before, but which are less sensitive. Rationality, originally seen as the means to open up the intuition to aspects of life outside the designer's experience, became, almost overnight, a tool-kit of rigid methods that obliged designers and planners to act like machines, deaf to every human cry and incapable of laughter,

says Jones. And he adds, almost as a footnote: 'That's what made me leave designing.'

Jones' personal response — which raises many questions of its own — has been to go to the opposite end of the process, to become a 'non-designer' who sees his job as creating the channels through which others can express their ideas about what technical objects and products should look like. One approach is to get people to think about their role in the design process in a new way; borrowing strategies from the American musician John Cage, here Jones suggests the need to develop situations analogous to Cage's silent music, by which an audience's self-awareness is stimulated by listening (or trying to listen) to a pianist who never plays a note. Another is to encourage a more explicitly political approach to the design process, which Jones shows to be the key point at which technological decisions are open to social input. 'Creative collaboration' is the desirable goal; 'attentiveness to context, not to self-expression, is the skill we have to foster, to encourage, to share ... the context, not the boss, has to become the manager of what is done, and how.'

The main thrust of his argument is that in the late 1960s and early 1970s, although many contemporary designers shared his feelings about the need to change the processes and techniques of designing to take social factors into consideration, they did not see the need to change its aims.

> We retained the concept of 'product' as the outcome of designing. We did not see that we were accepting only part of the challenge which we took up: the challenge to transform the idea of progress, which presumes a specific goal, into the idea of process, which does not. This transformation is, I now realise, a main event of the twentieth century, though it may have started earlier.

Should Modernism Rule?

It is at this point that Jones meets up with Berman. For the awareness of 'process' described by Jones is, in many ways, what Berman is describing as the spirit of modernity, a state of mind that emphasizes awareness of the present, and the potential for change that the present contains. Berman's book opens with the bold proclamation:

> There is a mode of vital experience — experience of space and time, of the self and others, of life's possibilities and perils — that is shared by men and women all over the world today. I will call this body of experience 'modernity' ... to be modern is to be part of a universe in which, as Marx said, 'all that is solid melts into air' — hence the title.

Berman, however, goes considerably beyond Jones in suggesting how this pursuit of modernity relates to political programmes. For Jones, despite his appeals for a new concept of the design process, remains primarily within the conceptual framework that determined much of the intellectual critique of the new left in the late 1960s and early 1970s (for example in criticizing the sterility and alienation of scientific rationality applied to social processes). Berman, however, turns sharply on this critique to suggest that it, too, suffers from a fatal flaw of one-dimensionality, the failure to grasp — in a way that I have already suggested above — the full meaning that the modern world takes on in the consciousness of the individual. He claims that in Marcuse's writing, for example, the masses 'have no egos, no ids . . . no dreams, no aspirations that are not given to them'. For others, he suggests, the critique of modernity has become an attempt to escape the challenge of the real world, not to come to terms with them; even worse, structuralism, he suggests, 'simply wipes the question of modernity off the map'. As for the 'pop modernists' such as Jones' hero John Cage, Berman says that he admires their creativity, but claims that they have never successfully captured the critical bite of their nineteenth century predecessors, and have been too easily co-opted by their sponsors.

> When a creative spirit like John Cage accepted the support of the Shah of Iran and performed modernist spectacles a few miles from where political prisoners shrieked and died, the failure of modern indignation was not his alone. The trouble was that pop modernism never developed a critical perspective which might have clarified the point where openness to the modern world has got to stop, and the point where the modern artist needs to see and to say that some of the power of this world has got to go.

These are some harsh judgements. And Berman's arguments have, unsurprisingly, already raised considerable controversy on the left. Perry Anderson, for example, in a long and mostly critical article in *New Left Review* ('Modernity and Revolution', *New Left Review*, No. 144, March/April 1984) claims that Berman ignores the specific political conjunctures under which enthusiasm for modernity as a cultural style was encouraged, and thus fails to see it as an integral component of a predominantly bourgeois ideology. The same point might be made with regard to Berman's sympathy for almost all of those who have spoken enthusiastically about the promises of science; readers of the *Radical Science* series may have difficulty in giving credence to a work that, at times, comes close to celebrating the romance of modern science and technology and the power over nature which it provides — a celebration that has long been criticized in these pages.

Yet Berman, I suggest, does have much to offer the radical science movement. When he writes that 'we don't know how to use our modernism; we have missed the connection between our culture and our lives', he is addressing a dilemma familiar to all those who have struggled with how to construct a concrete programme for locating science firmly in its social and political setting. We can also share his criticisms of successors to the Italian futurist movement, who placed the romance of the machine above all other human values, and his argument that 'the problem of all modernisms in the futurist tradition is that, with brilliant machines and mechanical systems playing all the leading roles... there is precious little for man to do'.

Berman addresses the problems of science most directly in a long chapter on the significance of Goethe's *Faust*. The scientific community has, in recent years, frequently been criticized for acting as a modern Dr. Faustus, opening up new empires of instrumental knowledge but lacking the wisdom to use it safely. Berman claims that the 'concerned scientists' of the post-war era 'established a distinctly Faustian style of science and technology, driven by guilt and care, by anguish and contradiction'.

Yet, he argues, Goethe's conception of Faust was not of a man riddled with doubts. Rather, here was a man searching for meaningful expression of his powers and finding it, in the climax of the verse-drama, in directing a vast operation to drain a large area of open marshland — work carried out by thousands of men carrying out Faust's instructions. His downfall comes when he orders the destruction of a small cottage that is getting in the way of the development work, an event in which the inhabitants of the cottage are unwittingly killed — a metaphor, suggests Berman, that epitomizes the 'creative destruction' that is frequently the price paid for human progress (and the real source of the torment that leads to Faust's downfall). If Faust embodied the spirit of modernism, then Goethe, according to Berman's characterization, was one of the first to realize the social dilemmas that this modernism was inevitably forced to face. And nowhere have these dilemmas been faced more starkly than when society has to confront the costs of its science-based technologies, ranging from a badly controlled pesticide factory in India to the world's armoury of nuclear weapons.

Many have been tempted to see in Faust little more than the model of the capitalist entrepreneur, the man ready to take risks with people's lives, including his own, and to sell his conscience for the sake of power over others. Berman suggests that Goethe may have had another concept in mind, a Faust who realizes in the end that mankind's progress will only come about through massive, collaborative endeavours, and that this can only be achieved through the large-scale organization of labour. But that this in turn, if it is to be properly directed, will itself require prodigious efforts of both social and political organization. It is, he claims, an argument that applies at almost all levels of social change — including even the debate, dear to Jones' heart, about the need for alternative/appropriate technologies:

> What seems to be ... important is to point out the intellectual vacuum that emerges when Faust is removed from the scene. The various advocates of solar, wind and water power, of small and decentralised sources of energy, of 'intermediate technologies' of the 'steady state economy' are virtually all enemies of large-scale planning, of scientific research, of technological innovation, of complex organisation. And yet, in order for any of their visions or plans to be actually adopted by any substantial number of people, the most radical redistribution of economic and political power would have to take place. And even this — which would mean the dissolution of General Motors, Exxon, Con Edison and their peers, and the redistribution of all their resources to the people — would be only a prelude to the most extensive and staggeringly complex reorganisation of the whole fabric of everyday life.

The challenge that Berman lays down for the radical science movement is to explore ways in which we can recapture the spirit of modernism expressed by Goethe's Faust without allowing this spirit to be shackled by the limited vision of capitalism. The ball does not rest, he suggests, with scientists, but with the wider community; for roles have been switched, and each of us is now his or her own Faustus — with scientists playing the role of Mephistopheles, the one who offers the deal.

> If scientific and technological cadres have accumulated vast powers in modern society, it is only because their visions and values have echoed, amplified and realised our own. They have only created means to fulfill ends embraced by the modern public: open-ended development of self and society, incessant transformation of the whole inner and outer world. As members of modern society, we are responsible for the directions in which we develop, for our goals and achievements, for their human costs.

Berman quotes the remark by an observer at the original atomic bomb tests in New Mexico that 'Good God . . . the long-haired boys have lost control.' But, he says, 'our society will never be able to control its eruptive "powers of the underworld" if it pretends that its scientists are the only ones out of control. One of the basic facts of modern life is that we are all "long-haired boys" today.'

According to Berman, many of the radical social and intellectual movements of the past two decades have helped to destroy the important promise that the modernist movement had previously offered, even to those on the left, by sketching out a broad vista of political opportunities. 'Virtually no-one today seems to want to make the large human connections that the idea of modernity entails,' he complains. 'The eclipse of the problem of modernity in the 1970s has meant the destruction of a vital form of public space. It has hastened the disintegration of our world into an aggregation of private material and spiritual interest groups, living in windowless monads', a fate in which he includes many of the most well-known intellectual figures of this period, including Foucault, Derrida, Barthes and Susan Sontag. 'In this bleak context, I want to bring the dynamic and dialectical modernism of the nineteenth century alive again,' says Berman. And science and technology have a key role in this process, indeed, may offer modernism the chance for social (as well as cultural) fulfilment which it has so far been denied. 'Modern machines have changed a great deal in the years between the nineteenth century modernists and ourselves; but modern man and woman, as Marx and Nietzche and Baudelaire and Dostoevsky saw them then, may only now be fully coming into their own.'

The Modernist Dilemma

There is much that could be criticized in Berman's argument. At times, for example, his enthusiasm for the modernist spirit seems excessive, tending to blind him to some of the less acceptable aspects of its social implications. Admittedly he is sensitive to this danger, illustrated most graphically in a detailed description of the way that the Faustian vision of the American town planner Robert Moses led to the brutal destruction of parts of Berman's native Bronx in order to make it easier for cars to gain access to New York. Elsewhere, however, such sensitivity is less obvious. There is little discussion, for example, of the alienation caused by reducing individuals to little more than minor, replaceable cogs in a vast productive machine, whatever the ideology — capitalist, socialist or Faustian — by which this productive machine is legitimated. Nor of the fact that many of the risks which, he suggests, are an inevitable part of any modernist project can themselves alter- natively be seen as socially determined. (See for example Dave Rosenfeld's analysis of the political root of the distinction between 'safe' and 'dangerous' practices in the nuclear industry, as described in *No Clear Reason*, Radical Science 14.)

Some of this confusion stems from the dilemma, already referred to earlier in this essay, that the supposed neutrality of science and technology is itself an integral part of the modernist project. The dialectic between the arts and the sciences is accepted — even by Berman — as an essential component of the modernist world-view. Injecting the idea of 'science and technology as ideology' into the argument therefore forces a certain amount of rethinking. Here Jones seems to have more to offer to Berman. By helping to decipher the social relations that underlie the conceptual assumptions of industrial design, and thus subsequently become embedded in the technical products that emerge from the design process, Jones provides us with important new perspectives on the phrase from Marx which Berman uses as the title of his book, perspectives to which Berman himself pays little attention. Thus even though Berman addresses the modernist dilemma head-on — that to create the future it is necessary to destroy the past — he gives little consideration to the particularly acute form in which this dilemma is faced by the radical science movement, namely, that the tools needed to construct this future tend to carry with them the imprint of the social relations that characterize their origins in this past.

Here too, however, Jones is less convincing. His book ends with a fictional dialogue on the design of an experimental city using new technology. There are two main characters, Utopia (described as 'the voice of perfection') and Numeroso ('the voice of all of us'). Spurred by a third character, Unesco, 'to rethink the possibilities [offered by modern technology] in a less restricted context,' Numeroso present an idealized shopping list of proposals:

> Education, largely without books, perhaps without literacy, even. The paperless office. No jobs. Everything is to be produced automatically, without workers or management, or else is made by hand by those who need it. No cash, just credit cards, and a guaranteed unearned income . . . Public access to all information, personal files included, no copyrights, no privacy of information, public or private... Private cars and buses to be replaced by automated minicars and minibuses accessible by credit card, and free of accidents, congestion, and parking problems, being computer controlled . . . Referenda by tv viewdata systems, every day . . .

and so on. Asked by Unesco about the development strategy which they would recommend to Third World countries, Utopia replies that it should begin

> with the most advanced technologies we know, but with a new factor that changes the picture completely: negative growth, or rather planned economic decline. The aim of the imaginary economy is to gradually approach the levels of income that are common in the majority of the world, a fraction of western incomes now. Why not try to close the gap from the top down? That's the idea we'd most like to see tried.

Admirable objectives, perhaps. And Jones does not fall into the trap of offering blueprints with no political dimension to them. Numeroso ends the dialogue by admitting 'where it's a proposal that seems to call for *persuasion*, shouldn't we be thinking of letting the inhabitants of the city decide for themselves? I don't think I understand the politics of this experiment. Who's in control?' Yet Jones does not present us with an agenda that points towards any direct form of intervention into the main political struggles of the 1980s (as opposed to those of the 1970s). Suggestions for 'public access to all information' or for 'planned economic decline' carry a utopian ring that has lost much of its charm as a guide to strategy in a period of increased militarization of Western economies and growing mass unemployment. To this extent, Jones seems to remain trapped within the post-modernist paradigm applied to technology, as I have described it above. It is not that he fails to grasp the relationship between social change and technical change, but that he displays a limited vision of power relationships that puts him more in sympathy with utopian cultural radicals such as John Cage than genuine political movements.

Berman, in contrast, does not offer an agenda but a form of inspiration. In doing so, he comes closer to the current political preoccupations of the left, particularly insofar as these relate to the social implications of rapid technological change. I write these words as the French Socialist Party is holding a major conference on the theme of 'modernization', addressing the results of a questionnaire distributed among rank-and-file activists which revealed that the 'idea of modernization is accepted by the great majority' of militants as 'an incontestable necessity, despite its brutality'. Meanwhile Martin Jacques, the editor of *Marxism Today*, is being quoted as telling the British Communist Party that 'the Left can appear conservative, for the past and against the future . . . [but] it won't appeal to the new generations'.

I also write at a time when the Commission of the European Economic Communities in Brussels is cutting back funding of research projects into nuclear safety and renewable sources of energy in order to pay for an industry-led programme of research into microelectronics; when Mr. Ian MacGregor of Britain's coal strike fame is seeking political support for building a new transportation infrastructure linking Europe's capitals with high-speed trains and a channel tunnel at the same time as he is deliberately eliminating tens of thousands of jobs in the coal industry; when Mrs. Margaret Thatcher is considering committing hundreds of millions of pounds to the construction of a space station while hospitals complain of the lack of even basic medical facilities; or when the French Socialist Party is continuing to suppress internal opposition to its nuclear programmes — both civilian and military — on the grounds that such opposition would weaken France's bargaining power in international relations.

Berman himself does not address any issues like this directly. But what he does do, I think, is show how they must be accommodated in the agenda of left criticism of technology, a task which is necessary if our arguments are to remain relevant to left strategy more generally. Three particular themes have tended to dominate this agenda in the recent past: technology as a source of human oppression, exploitation and alienation (the dominant theme in Marx's own writings on the negative aspects of technical change); technology as the principal means by which mankind has exploited the natural environment; and finally, technology as the expression of hierarchical social relations. Berman suggests that the time has come to reinject a further dimension, technology as a source of human fulfilment that embraces both aesthetic experience and collective social action. Integrating this theme into the previous three is not likely to be straightforward (particularly in cases where it challenges previous assumptions, such as the Marcusian notion of science and technology as the apparently *neutral* legitimation of political power). But it may prove essential if the radical science movement is to find a path out of the modernist dilemma that it now faces.

Bibliographic Note

This essay was occasioned by a reading of the first two books below, in light of the third.

Marshall Berman, *All that is Solid Melts into Air*, New Left Books, 1984.

J. Christopher Jones, *Essays in Design*, Chichester and New York, John Wiley, 1984.

A.N. Whitehead, *Science and the Modern World*, CUP, 1925, republished by Free Association Books, 1985.

A further discussion of the 'post-modernism' debate can be found in Fredric Jameson's Foreword to Jean-François Lyotard, *The Postmodern Condition: A Report on Knowledge*, Manchester University Press, 1984, pp. vii–xxi. Lyotard's book was first published in France in 1980 and presents a somewhat different perspective from that of Habermas. The latter's ideas are expressed most concisely in 'Modernity vs. Postmodernity', *New German Critique* 22 (Winter 1981), pp. 3–14, as part of a special issue on modernism; this is a printed version of a lecture delivered in September 1980, when Habermas was awarded the Theodore Adorno prize by the city of Frankfurt. For a reply to Habermas, see other articles in the same issue, especially the one by Anthony Giddens, 'Modernism and Postmodernism'.

LICENSING REPRODUCTIVE TECHNOLOGIES?

Edward Yoxen

Essay review of Peter Singer and Deane Wells. *The Reproductive Revolution: New Ways of Making Babies*, Oxford University Press, 1984, Pp. 273, £2.95; and Jonathan Glover, *What Sort of People Should There Be? Genetic Engineering, Brain Control and their Impact on our Future World*, Harmondsworth, Penguin, 1984, Pp. 190, £2.50.

In the closing weeks of 1984, it became clear that there was an intense opposition to research in human embryology and to the practice of *in vitro* fertilization (ivf) from extreme right-wing MPs and the anti-abortion lobby in the UK. As I write, in January 1985, the present controversy — over the latest case of surrogate motherhood to be publicly revealed — suggests that this year, which is likely also to see another attempt to carry out gene therapy with human patients, will be one of mounting political conflict over novel reproductive and genetic technologies and over the moral, legal and social acceptability of different forms of procreation, parenting and medical care. The ideological tensions surrounding ivf and genetic engineering have finally erupted into the national, parliamentary political arena, fifteen years after Robert Edwards achieved *in vitro* fertilization of human eggs. It is likely therefore that there will now be a more serious and intense public debate on some of the issues in this area, most of which could have been tackled years ago, centred around the question of which procedures and experimental programmes should be licensed and given resources, if any.

Although these are important matters, I would argue that the politics of reproductive technology should be construed much more widely, to include such questions as how social and medical needs originate, how medical procedures and technologies (such as ivf) are constituted and what counts as expertise. In the face of imminent controversy, one approach seeks merely to license selected areas of work; such an approach takes the technology and its institutional setting as given and unproblematic, and ignores these constitutive questions in the pragmatic search for a workable consensus. In the present context, therefore, books which might broaden our perspective on and critical appreciation of genetic and reproductive technologies are provisionally to be welcomed. On the face of it *The Reproduction Revolution* and *What Sort of People Should There Be?* ought to fulfil exactly that role. I believe the former is helpful in this way, the latter not.

These two books are works in applied philosophy. In their rather different ways they exemplify a welcome concern with concrete problems. They are part of a recent cultural phenomenon of academic philosophers trying to add something to debates about the social relations of science, medicine and technology. The book by Singer and Wells is certainly informative and adds to our understanding of ivf; that by Glover is far more abstract and esoteric, so that only a few readers, I imagine, will find it intellectually engaging. But both books neglect the forces that drive research forward or that inhibit more progressive developments. In this respect they indicate the limitations of a philosophically-direct analysis of science that lacks a political basis. It might be more accurate to speak of a meditation on science in Glover's case.

The Reproductive Revolution

Peter Singer is Professor of Philosophy and Director of the Centre for Human Bioethics at Monash University. He is the author of several books on ethics and is known particularly for his advocacy of animal rights. Deane Wells is a member of the Australian Parliament. They deal with practical questions, such as legislation, codes of professional practice, hospital policy and social mores. Their principal focus is on ivf; they move on to surrogate motherhood, they touch on cloning and end with gene therapy. Their book is topical, technically informed and comprehensive, but not excessively detailed. Theirs is essentially an exercise in moral reasoning. Even though I think their treatment has its shortcomings. *The Reproduction Revolution* is a fine piece of serious, popular writing.

They begin by describing the genesis of Edwards' and Steptoe's collaborative work on *in vitro* fertilisation, and show the relative ease with which fertilization was achieved, after a year's research, compared with the difficulty of bringing about implantation and a complete pregnancy, after seven more years. Singer and Wells contrast an eventual success for Jan Brennan with an eventual resignation to failure for Isobel Bainbridge. I wondered whether a 'balanced' treatment ought to have included three such failures to the birth of one baby. Certainly both accounts show something of the psychological struggles for these women and their partners in seeking and receiving medical assistance with involuntary infertility. The point is not emphasized, but it is there to see, that both women suffered through early medical incompetence, regardless of the 'heroic' subsequent efforts made to help them. It is also interesting that they mention Mrs. Bainbridge's feeling of having gained something from her struggle with infertility. She is not left as a passive victim of her circumstances. The introduction continues with some thoughtful comments on success rates, where the question is which figure you choose — pregnancies/patient, pregnancies/treatment, pregnancies/egg fertilized or per egg transferred to the womb. Thus informed; I turned from this section to the data in the Warnock report, which describes figures supplied by Edwards and Steptoe from their work at Bourne Hall. These figures seem to be presented in such a way that meaningful success rates cannot be derived from them. One just has an impression of success, for some people. There are then some interesting data on research funding. Carl Wood's group in Melbourne received $1 million through the 1970s from American foundations, although they supported no work inside the United States. Why did the Australians do so well financially? The first chapter closes with a description of the opinion surveys, which show a high degree of popular support for ivf for infertile couples. To me all this represents an excellent introduction.

The next three chapters take up the implications of ivf, starting with the 'simple case', moving on to various forms of donation and the storage of and research on embryos, and thence to surrogacy. It is in the second chapter that the philosophical character of the book is made clear. In a way that is also apparent in the Warnock report, one has the sense of someone following the traditions of their discipline in developing a reasoned moral argument. Singer and Wells take issue particularly with natural law theologians — those who argue that the sacredness of *in vivo* fertilization is embodied in the fittedness of sexual intercourse to the essential biological goal of procreation. Indeed they turn the theological argument opposing ivf neatly on its head.

> Indeed we would say that those natural law doctrines which Catholic scholars have most enthusiastically embraced lead to the conclusion that IVF should be encouraged. Remember that it is not IVF that prevents sexual intercourse from achieving its essential goal of procreation. Accident or disease has already seen to that. IVF aims at restoring to the infertile person the possibility of procreation ... When people who desire children are prevented from reproducing, they are

prevented from achieving the form of fulfilment or flourishing that they seek. So at least to the extent that biotechnology serves as a therapy for unwanted infertility, natural law theorists ought not merely to permit it: they ought to be among its strongest advocates.

Later on in the same chapter they make the same move against the Catholic bishops who stressed the psychological risks to ivf children, who may receive excessive attention from the media. It takes Australian radicals to point out that the same would be true of royal children, although no-one has yet made that an argument for Republicanism.

One may wonder whether this kind of scoring of rationalist points against theologians has any great value. Certainly it would be true to say that such jabs are likely to have little effect on the attitudes of many devout but unreflective Christians, with very fixed ideas about the nature and significance of conception. But I suspect teleological arguments of this kind have a much wider currency than one supposes in underpinning notions of what is natural or acceptable. As part of the more general exercise of developing a humanist world-view, such philosophical skirmishing seems to me worthwhile.

Singer and Wells also tackle here the questions of who should bear the costs of ivf, or indeed whether it represents a misallocation of resources on luxury medicine. They offer some figures which show what *may be charged* for different forms of medical assistance in Australia. On this basis ivf is not phenomenally expensive at around A\$ 2,000 to A\$4,000 for each attempt, by comparison with costs, say, of tubal surgery. Nor is it life-threatening. Singer and Wells argue that it should be regarded as the correction of a major disability or as alleviating a likely cause of severe mental stress, and should therefore be supported by the state or by medical insurance schemes.

'Beyond the simple case' of helping infertile couples, using their gametes, without any storage, lie the questions of sperm, egg and embryo donation, assistance for lesbian couples or single women, and the storage of embryos. Their position on all these questions is relatively permissive. As one might expect, they argue against according embryos a special moral status, equivalent to that granted to people after birth, from the moment of conception. Their 'cut-off line' for embryo culture or experimentation is the onset of brain function, set provisionally at somewhere around the end of the sixth week after conception. This limit is set by deliberate analogy with the definition of death in terms of the cessation of brain function. It is much further into the development process than the limit of fourteen days proposed by the Warnock committee.

The next chapter takes up the question of surrogacy. They are prepared to accept 'full surrogacy', where the woman who carries an implanted embryo to term and hands the baby over shortly after birth to its genetic parents, having herself made no genetic contribution. They see this as likely to be less complicated than 'partial surrogacy' where the host or surrogate mother is also one of the genetic parents. They analyse some of the cases handled by the American lawyer, Noel Keane, and conclude that the law of contract is really inadequate to deal with the breakdown of surrogacy arrangements. They argue firstly for the banning of private, profit-making surrogacy agencies, like Surrogate Parenting Associates, Inc. based in Kentucky, and secondly (by analogy with adoption) that limited use of surrogacy be controlled by the state.

This now introduces the theme of 'licensing', of surrogacy, of ectogenesis (extra-uterine development), of single cloning and of gene therapy, which runs throughout the rest of the book. In the final chapter they argue for a kind of standing commission to monitor all these issues. When compared with the first three chapters, the second half of the book lacks thoughtfulness. They suggest that sex-selection would conduce to family limitation and would therefore be justified as helping to check the population explosion. They also ignore the arguments against developing gene therapy on the grounds that antenatal diagnosis can be shown to be successful already as a mode of prevention and as one that is acceptable to most people. It is clear that they have not researched the genetic issues as thoroughly as was the case with ivf. They converge on Glover's abstract endorsement of regulated genetic engineering, in which one would begin with the amelioration of genetic disease and move on eventually to positive eugenics.

The notion of licensing scientific research and medical procedures is basic to the approach. They consider linked areas of activity and show that some kinds of objection to them have no logical or moral force. Though the two books are otherwise composed rather differently, both *The Reproduction Revolution* and *What Sort of People Should There Be?* have this analytic strategy in common. Singer's and Wells' book is also very like the Warnock report in approach, even if their conclusions diverge. They are written as advice, taking certain practices as given or established, and offering ways of revising, limiting and licensing them.

One effect of this approach is to ignore the question of why things have come to be this way. Thus academic and professional motivations, medical status hierarchies and priorities, and people's expectations of medical expertise go unexamined. It would be idle to criticize the Warnock committee for not having done this, since that was not their remit, but Singer and Wells could have done. One might ask what set of career choices led Edwards and Steptoe into their collaboration, or why their interpretation of the risk studies in animals was much less cautious than anyone else's, or why there are so few data on the effects of the drugs and invasive procedures used in ivf, or why there are so few data on the causes of infertility or why the waiting lists for ivf clinics are so long when the success rates are low. Singer and Wells have been asking: Is ivf, in all its forms, a legitimate solution to some of the problems of infertility? My question, which does not pre-empt or negate theirs, would be: What are the factors which make us think of ivf as a solution anyway? In this respect I think the Warnock report is somewhat more enlightened than *The Reproduction Revolution*. The report does at least begin with a discussion of infertility services and identifies a few of their shortcomings.

Genetic Supermarket

Jonathan Glover's book is really rather different. Its most obvious feature, and one that clearly distinguishes it from *The Reproduction Revolution*, is its studied abstraction in the face of deep popular concern. It is not that I think Glover is not intellectually serious; it is just that I find his way of being serious completely alien, and I am sure that others will too. Basically I found this book a disappointment, since I share Dr. Glover's view that discovering our prejudices, anxieties and false beliefs about imminent technological, cultural and political developments is worthwhile, and that articulating a coherent set of values for the design of new technologies is an important social exercise. Philosophy is a resource for thinking about the future, amongst other things, and it is one that we have to be taught how to use. Holding up common sense about human nature to the light of reason is something more people ought to do. But the light can get in your eyes. It does here.

The first third of the book is about genetic engineering, but not the forms going on now — genetic screening of adults and adolescents and antenatal diagnosis — nor about the forms that may be attempted soon — gene therapy at some point after birth. Instead Glover considers the more remote prospect of genetic engineering to create preferred types of individuals. Having dispenses with some wrong-headed objections to the theoretical feasibility of such actions, which some people have used as an excuse for not thinking any further, he tackles head-on the issue of parental or societal selection of particular genotypes, to ask on what basis it might be acceptable. In so doing he knowingly bypasses real and more pressing problems. For example, plans to screen schoolchildren for particular conditions hold the real prospect of stigmatization, racial discrimination and anxiety as a consequence. There are also plans to correct genetic defects by supplying to specific tissues packages of genes that are lacking in some individuals — plans which are being readied now in the USA and may be implemented this year. I have to say that I think it takes real perversity to bypass these problems, even if there are 'deeper' philosophical questions raised by developments in a more distant future. If I was being malicious, I would say that this amounted to a form of intellectual dandyism, that should be criticized as such. But Dr. Glover comes across as just too well-meaning and sincere for that aggressiveness to seem entirely fair.

As it is, Dr. Glover endorses a form of what has been called 'the genetic supermarket', which would allow parents a degree of freedom to exercise genetic preferences whilst attempting to keep away from governments and social elites the ability to manage or impose such preferences. Although I have deep misgivings about state agencies intervening in the market of genetic preferences — and this incidentally is a feeling that Singer and Wells seem to lack in their enthusiasm for State Surrogacy Boards and so on — I find the supermarket unconvincing and unappealing as an image of freedom. After all, we enter supermarkets with carefully nurtured preferences and select from a restricted set of packaged commodities. Rather, I would see the supermarket as an image of highly constrained choice in commodity terms. Speaking in this consumerist way implies that one is thinking of individual propensities and qualities as if they were fixed and purchasable. For me this is investing characteristics and abilities, assumed to have a genetic basis, with a misplaced concreteness. The value of this opening section of the book is debatable.

The underlying message is clear: Don't condemn genetic engineering *just because* it is genetic engineering, since it *could* be a means to a better existence. And of course thinking about possible futures now does actually get us to bring to the surface and affirm important values, relevant to living our lives now. Various prejudices against genetic engineering are revealed as such, as the philosophical scalpel slices away layers of irrational resistance and misconception. But is that really the most important didactic point to make at this moment in time? Some people will, I think, be stunned by the breathtaking leap into an indefinite future; others, more accepting of the humane and witty meditation on the kinds of values appropriate for long-term, transgenerational modification of human nature. It all depends whether you like the word-games that philosophers play.

In the brain control section, which is the bulk of the book, the didactic method changes to a series of thought experiments. This involves discussion of a series of increasingly complex hypothetical machines. There is a thought-reading machine, an experience machine, a dream-world machine and a horror machine; and very subtle they are, too. Halfway through I suddenly had the paranoid suspicion that all this was a very indirect, cryptic commentary on the social impact of holography, personal stereo tape-recorders, video-recorders and intelligent knowledge-based systems, written by a master caricaturist. At one level I suppose it is that. Then again it could be the result of an attempt to cope with guilt about liking science fiction. In parts it does degenerate to making elaborate jokes about philosophers that few people will find funny.

But the overall message is again reasonably clear: modifications to human nature and mental capacities — this time by something other than genetic engineering — are not necessarily unacceptable; they should be based on a number of general principles, including the importance of active self-development, rather than the passive consumption of experience, the need for social contact and variety and the value of the continuing development of consciousness and its modalities. These are noble sentiments to affirm, but Dr. Glover takes the most elaborate route via these technologies of fantasy to reach them. Following him across the high-level traverse to the ledge on which these thoughts are placed is hard work. Reading Chapter 10, 'Thoughts on the Thought Experiments', where he justifies his abstraction, is like the welcome relief after a long hard climb, before the ascent to the summit in the chapters on obligations to future generations. To put that another way, Dr. Glover uses a variety of bizarre devices to make clear something of the subtlety and complexity of human mental processes, experience and individuality, but you have to be *really* interested in the mind to put up with such an extraordinarily elaborate performance. Or you may just find it all hilarious.

The first machine is a device that makes thoughts 'transparent' and displays them on a screen where other people can see them. Imagine if you like a kind of breast-plate on which what one 'really' thinks is projected. Assuming that we did not all learn to avert our gaze, this would be an obvious threat to privacy as we think of it at present. Thinking through the psychological implications of such a device allows Dr. Glover to say some interesting things about the need for a degree of self-concealment or evasiveness in the development of individuality and about the possible value of a greater openness to others. Frankly this is a bit like a rather precise vicar discussing nudism. Then he considers mood control; and opting again for a technology that individuals could choose to use, he discusses how a 'superdrug' for developing emotional discrimination and robustness could be acceptable.

The next chapter concerns the control of unwanted behavioural impulses by benevolent or malign authorities, which is quickly condemned, or by sensible citizens exploring their own natures. Highly convergent, single-minded people might well use them to concentrate their energies and divergent thinkers with multiple interests might do exactly the opposite. Interesting, I suppose. I hope to convey here something of the flavour of the book or at least the nature and tone of the conclusions. Technology *can* be life-enhancing, particularly if already sensitive and responsible people use it to develop a set of capacities for empathy, sensitivity and creativity. An eminently reasonable thing to say, but verging on the platitudinous, don't you think? What I can't convey here is the amount of boiling to which I had to subject Dr. Glover's book to distil these simple ideas out of it. Of course in so doing I have denatured some parts of the argument but I do wonder if it was worth the cost of the fuel.

The next two chapters are various kinds of fantasy-realizing devices, incredibly complex simulators where chosen experiences can be 'realized'. It takes two hard chapters to say that these are not a good idea. Then there is a sequel on the existential meaning of work in a society where automation has abolished the need for human productive labour. For me the point that real human interaction is more important in the development of individuality and responsibility was made far more powerfully by an article in a forensic science journal that I found by chance recently in a university library, and wished I had not. It showed the room in which a computer programmer had died, whilst connected up to a plastic female doll, watching a pornographic video on a screen at the foot of the bed and inhaling amyl nitrate through a respirator. This sad corpse spoke volumes about the tragic search for private experience, facilitated by new technologies.

The final section of the book deals with the problem of deciding on the values that should inform modifications of human nature that will influence the lives and potentiality of distant generations. How can we balance perceptions of advantage to ourselves against the fact of ignorance of what our descendants would wish us to have done. Various kinds of moral theory are tried out in this role but none of them appear very helpful. Everyone needs a powerful graphic image to keep one going here, and in this section it comes in the form of an allegory for nuclear power, a bus on which a powerful bomb has been placed, from which we can get down somewhere, hopefully before the bomb has gone off, hoping that our descendants will know how to defuse the problems or will not mind when it blows up. The implication is that our present attitudes to nuclear power — or the attitudes of those who support its further development — have that moral character and are therefore indefensible. The conclusion only has force if you accept the analogy in the first place.

My lasting impression of this book is a sad one. Surely there has to be a more appealing way of writing seriously and analytically about the interaction of technology with the intimate structure of our lives? I know that thought experiments are an intellectual's device for revealing the elusive in the familiar, but they are also an evasion. They translate all the moral dilemmas and political choices into an indefinite region in space and time where only a very few people can deal with them, only those that is who have learnt the rules of constructing word-games. And on that rarefied level all the passion and coarseness, the things that make moral choices hard to take in practice, have disappeared. They relate, I think, to a sort of trickle-down theory of moral sentiments, in which the ethically skilled allow their values to percolate to the rest of us, for the general good.

A Licensing Agency?

Here then are two thoughtful books, both the product of sustained intellectual work, even if Jonathan Glover has ploughed a rather strange furrow. In their respective ways, and obviously I prefer the 'Australian' way, they are trying to stimulate public debate about present technical change and about long-term questions such as the kinds of people we should strive to create in our society. And whatever one may feel about the style of Glover's book, he is at least trying to anticipate future dilemmas.

But public debate is a deeply ambiguous notion. It is often seen as an end in itself or as the moment of transient lay concern that makes necessary policy formation by governmental elites. In the months since the Warnock report appeared in Britain, helped perhaps by the curiously large number of leaks of its contents, public discussion of ivf has increased. A great deal of the discussion has been about experimentation on embryos and surrogate motherhood, particularly the former. Most notably this was the case in the House of Commons debate on the report. As Mary Warnock has said herself there is a great danger that the other recommendations, for example those relating to the infertility services, will simply get ignored in all the sound and fury about the licensing or prohibition of embryo research. That would be a shame, no doubt, but I do not think that gets to the deeper issues of how reproduction, fertility, infertility, pregnancy and sexuality are handled in our culture. Much of the practice of reproductive medicine, particularly that concerned with infertility, is oppressive, humiliating and anxiety-creating for those seeking assistance. Licensing new reproductive technologies will not remove these problems: indeed it is likely to embed them more effectively in the social relations of contemporary medicine.

Some kind of regulation is necessary for ivf, particularly to forestall the predation of private medical entrepreneurs on anguished would-be parents, but licensing of this kind always seems to treat as quite unproblematic the concomitant strengthening of the power of senior members of the medical profession to act behind the smokescreen of professional integrity. Both the Warnock report and *The Reproduction Revolution* totally ignore this issue.

At least as important as regulations and legislation, overseen perhaps by a committee or licensing authority, is a campaign to research the medical, biological and cultural basis of ivf and other contemporary reproductive technologies. I have in mind work on the social biology of infertility to consider whether existing, naturalized social practices conduce to infertility. Areas that need investigation include the acceptance of misdiagnosis of abdominal pain as dysmenorrhoea, hospital wound infection rates, unnecessary gynaecological surgery and the cumulative effects of ignorance of basic human biology compounded by a fear of sexuality. Similarly ivf clinics have enormous waiting lists, despite the relatively low chance of success, and the frequent experience of humiliation in seeking assistance with infertility. How demand for ivf is constituted, so as to make it seem a 'solution' — which indeed it will be, for a few people — needs critical examination. So does the long-term effect of ivf investigation. It is not at all impossible that the clinical enthusiasms of today, using powerful hormonal preparations and repeated investigation of the ovaries, will generate unexpected problems in the decades to come. What is required to shift the balance of power over the development of reproductive technologies is not a politics of elite control through licensing and negotiation, but a politics of constant lay appraisal.

CONNECTIONS

PUBLICATIONS RECEIVED AND NOTICED
Published in London unless otherwise noted.

Huw Beynon, ed., *Digging Deeper: Issues in the Miners' Strike*, Verso, 1985, Pp. xiv + 252, hb £16.50, pb £3.95.

Ruth Bleier, *Science and Gender: A Critique of Biology and Its Theories on Women*, Oxford, Pergamon Press, 1984, Pp. xii + 219, £8.35.

Renate Bridenthal, Atina Grossman and Marion Kaplan, eds., *When Biology Became Destiny: Women in Weimar and Nazi Germany*, New York, Monthly Review Press, 1984, Pp. xiv + 364, pb $12.00.

W.F. Bynum, E.J. Browne, Roy Porter, eds., *Dictionary of the History of Science*, Macmillan, 1983, Pp. xxiv + 494, pb £6.95.

Sam Cole and Ian Miles, *Worlds Apart: Technology and North-South Relations in the Global Economy*, Brighton, Wheatsheaf Books, 1984, Pp. xviii + 283, £25.00.

Peter Collier and David Horowitz, *The Kennedys*, Secker and Warburg, 1984, Pp. 576, £12.50.

Kenneth C. Davis, *Two-Bit Culture: The Paperbacking of America*, Boston, Houghton Mifflin Co., 1984, Pp. xvi + 430, pb $9.70.

William C. Dowling, *Jameson, Althusser, Marx: An Introduction to the Political Unconscious*, Methuen, 1984, Pp. 152, pb £4.95.

D. English, B. Epstein, B. Haber, J. Maclean, 'The Impasse of Socialist-Feminism', *Socialist Review* 79 (1985) 93–110.

Wendy Faulkner and Erik Arnold, eds., *Smothered by Invention: Technology in Women's Lives*, Pluto Press, 1985, Pp. 276, pb £7.95.

Ray Hammond, *The Writer and the Word Processor: A Guide for Authors, Journalists, Poets and Playwrights*, Coronet Books, Hodder & Stoughton, 1984, Pp. 224, pb £2.95.

Paul Heyer, *Nature, Human Nature, and Society: Marx, Darwin, Biology and the Human Sciences*, Greenwood Press, 1982, Pp. xvi + 266, £33.95.

A.R. Jones, 'Julia Kristeva on Femininity: The Limits of a Semiotic Politics', *Feminist Review*, 18 (1984) 56–73.

R.D. Laing, *Wisdom, Madness & Folly: The Making of a Psychiatrist 1927–1957*, Macmillan, 1985, Pp. x + 146, £9.95.

Peter Large, *The Micro Revolution Revisited*, Frances Pinter, 1984, Pp. vii + 216, hb £9.95, pb £4.95.

Larry Laudan, *Science and Values: The Aims of Science and Their Role in Scientific Debate*, University of California Press, 1985, Pp. xiv + 149, £13.75.

George Levine and U.C. Knoepflmacher, eds., *The Endurance of Frankenstein: Essays on Mary Shelley's Novel*, University of California Press, 1979, Pp. xx + 341.

Norman Myers, ed., *The Gaia Atlas of Planet Management*, Pan Books, 1985, Pp. 272, pb £9.95.

Redmond O'Hanlon, *Joseph Conrad and Charles Darwin: The Influence of Scientific Thought on Conrad's Fiction*, Edinburgh, The Salamander Press, 1984, Pp. 189, £17.50.

Marge Piercy, *Vida,* Penguin, 1980, Pp. 448, pb £2.50.
David Pilgrim, ed., *Psychology and Psychotherapy: Current Trends and Issues,* Routledge & Kegan Paul, 1983, Pp. x + 236, pb £7.95.
Radical Philosophy 37 Summer 1984, £1.50. Special Issue: Science, History and Philosophy.
Diana Ralph, *Work and Madness: The Rise of Community Psychiatry,* Montreal, Black Rose Books, 1983, Pp. 216, pb.
Julia Segal, *Phantasy in Everyday Life: A Psychoanalytical Approach to Understanding Ourselves,* Penguin, 1985, Pp. 234, pb £2.95.
Dale Spender, *Women of Ideas (and What Men Have Done to Them): From Aphra Behn to Adrienne Rich,* Ark, 1983, Pp. x + 800, £4.95.
Brian Stableford, *Future Man,* Granada, 1984, Pp. 192, £9.95.
Bryan S. Turner, *The Body & Society: Explorations in Social Theory,* Oxford, Basil Blackwell, 1984, Pp. vii + 272, £16.00.
Hebe M.C. Vessuri, *Ciencia Academica en la Venezuela Moderna: Historia Reciente y Perspectivas de las Disciplinas Cientificas,* Caracas, Acta Cientifica Venezolana, 1984, Pp. 461, pb.
Martha S. Vogeler, *Frederic Harrison, The Vocations of a Positivist,* Oxford, Clarendon Press, 1984, Pp. xvii + 493, £27.50.
Alfred North Whitehead, *Science and the Modern World* (1926), Free Association Books, 1985, Pp. xxiii + 265, hb £11.95, pb £4.95. A reprint of a classic text, with a new Foreword by Robert M. Young and a new bibliography.
Richard Wollheim, *The Thread of Life,* Cambridge University Press, 1984, Pp. xv + 228, £20.00.

INTERNATIONAL MEETING OF RADICAL SCIENCE PERIODICALS

The 1985 Easter meeting, held at the Free Association Books premises in London, was a smaller-scale, more informal event that in previous years. Attending were members of the collectives that produce *Wechsel Wirkung, Naturkampen, Revoluon, Science for People,* the Radical Science Series and a proposed new UK magazine, *Potentials.* Instead of a public conference, we arranged simply an enlarged meeting on the Friday, where we discussed two main presentations: from a BSSRS group on their book[1] warning about new police technologies, and from a group of anti-racist science teachers on their work.

On the Saturday morning we heard reports on the situation of each of the journals, some of them as precarious as ever. Sunday's discussions concerned mostly biotechnology — the weekend's main theme.[2] Gratefully accepting *Naturkampen's* offer to host the 1986 meeting in Denmark, we discussed the possibility of going public there with the theme of holistic medicine and new concepts of nature. Lastly, on the Monday morning, we concluded with a session on the economics of publishing and marketing — centring on the experience of Free Association Books, but also increasingly relevant to other participants as their journal collectives also move towards book publishing.

Notes
This report has been kept unusually brief for lack of space (L.L.).
1. BSSRS Technology of Political Control Group, *TechnoCop: New Police Technologies*, Free Association Books, 1985, £3.50.
2. Notes of the biotechnology discussion are available on request.

NOTES ON CONTRIBUTORS

Doug Boucher teaches biology at the Université de Québec à Montréal. He does research on agriculture in Québec and Nicaragua.

David Dickson is a science journalist who currently works for the US journal *Science*. A founding member of the *RSJ* editorial collective, he is the author of *Alternative Technology* (1974) and, more recently, *The New Politics of Science* (New York, 1984).

Dave Feickert works for the NUM in Sheffield.

Ludmilla Jordanova is Lecturer in History at the University of Essex. She writes about science, medicine and culture; she is currently working on a book concerning debates about the family, gender and sexuality in 18th century Britain and France.

Vincent Mosco is Professor of Sociology at Queen's University, Kingston, Canada. He is the author of *Pushbutton Fantasies: Critical Perspectives on Videotext and Information Technology* (Ablex, 1982).

Isadore Nabi is a distinguished scientist whom *Nature* has tried to liquidate (*Nature*, 3 September 1981).

Don Parson, having recently finished his Ph.D. in urban planning at UCLA, expects to publish his first book, entitled *Autonomy and Planning: Essays on Liberation and Planning Theory.*

Jonathan Rée teaches at Middlesex Polytechnic in London. His latest book, *Proletarian Philosophers*, was published by Oxford University Press last year.

Bruno Vitale should teach theoretical physics at the University of Naples but works with children at the Centre for Genetic Epistemology in Geneva. He is co-author of *Symmetry and Reduction of Dynamic Systems* (Wiley, in press) but is mostly interested in the role and impact of military-funded research on science policy in our countries.

Edward Yoxen teaches history and philosophy of biology and science policy at the University of Manchester. He is currently working on a book on *in vitro* fertilization and medical genetic engineering.

COMRADELY PUBLICATIONS AND GROUPS

(Prices are subject to change)

ALTERNATIVE PRESS INDEX
Alternative Press Centre, P.O. Box
7209, Baltimore, MD 21218, USA.
AMPO Japan-Asia Quarterly
Review
P.O. Box 5250, Tokyo International,
Japan. 4 issues p.a. for $16, $24 inst.
ANTIPODE-A Radical Journal of
Geography
P.O. Box 339, West Side Station,
Worcester, MA 01602, USA. 4 issues
for $12 employed, $8 unemployed,
foreign $2 extra.
ARENA-A Marxist Journal of
Criticism and Discussion
P.O. Box 18, North Carlton, Victoria
3054, Australia. 4 issues for A $8,
overseas A $14.
ASSOCIATION OF RADICAL
MIDWIVES
c/o 13 Fremont Street, London E9,
Tel. 986 8939.
BERKELEY JOURNAL OF
SOCIOLOGY
458A Barrows Hall, Dept. of
Sociology, Berkeley, CA 94720, USA.
Subscription $5 individual, $10 inst.
CAHIERS GALILEE
c/o G. Valenduc, 5 rue de la
Resistance, 1490 Court-St-Etienne,
Belgium, special issues on
biotechnology, infotech, 80 FB each.
CAPITAL & CLASS-Journal of the
Conference of Socialist Economists
25 Horsell Road, London N5.
Membership rates: £9, low income £6,
overseas £10/£6, for 3 issues p.a.
CIENCIA HOJE
Av. Venceslau Braz, 71 fundos casa
27, 22290 Rio de Janeiro–RJ, Brasil.
CINE-TRACTS-A Journal of Film
and Cultural Studies
Institute of Cinema Studies, 4227
Esplanade Ave., Montreal H2W,
Quebec, Canada. 4 issues p.a. for $10,
foreign $12, inst. $20.
COMMENT-Libertarian newsletter
published by Murray Bookchin
P.O. Box 371, Hoboken, NJ 07030,

USA. Send stamped sae.
COUNTER-INFORMATION
SERVICES (CIS)
9 Poland Street, London W1. The
Nuclear Disaster, 85p; The New
Technology, 95p.
CRIME AND SOCIAL JUSTICE
– A Journal of Radical Criminology
P.O. Box 40601, San Francisco, CA
94140, USA. 2 issues $8, $18 inst.;
foreign $10/$20.
CRITICAL SOCIAL POLICY
46 Elfort Road, London N5.
CRITIQUE-A Journal of Soviet
Studies and Socialist Theory
31 Cliveden Road, Glasgow G12
0PH.
CRITIQUE OF
ANTHROPOLOGY
Luna, PO Box 6004, 1005 EA
Amsterdam, Holland.
DEMOCRATIC PALESTINE
(formerly PFLP Bulletin)
Box 12144 Damascus, Syria. $15 p.a.
for 12 issues.
DESARROLLO-Tribuna para una
politica scientifico-Technologica
Apdo. 388, San Pedro 2050, Costa
Rica.
DIALOGO SOCIAL
Ediciones CCS, Apartado 9A-192,
Panama, R.P.
ETCETERA-Correspondencia de
la Guerra Social Apartado Correos
1.363, Barcelona, Spain. No. 5
(Febrero 85) on 'Technologia y
Sociedad'.
GLOBAL ELECTRONICS,
Pacific Studies Center, 222B View
Street, Mountain View, CA 94041,
USA. 12 issues p.a. for $15. Also
'Changing Role of S.E. Asian
Women', $2; 'Delicate Bonds: The
Global Semiconductor Industry', $2.
The GUARDIAN-Independent
Radical Newsweekly
33 West 17th St., New York, NY
10011.
HEAD & HAND-Socialist Review

of Books
CSE Books, 25 Horsell Road,
London N5.

HEALTH/PAC (Policy Advisory
Committee) BULLETIN
17 Murray St., New York, NY 10007.

HISTORY WORKSHOP-A
Journal of Socialist and Feminist
Historians
c/o Routledge & Kegan Paul plc,
Broadway House, Newtown Rd.,
Henley-on-Thames, Oxfordshire RG9
1EN. 2 issues p.a. for £10 ind., £15
inst.

INSURGENT SOCIOLOGIST
Dept. of Sociology, Univ. of Oregon,
Eugene, OR 97403, USA. 4 issues p.a.
for $20 individual, $10 low income,
$20 inst.

INTERNATIONAL JOURNAL OF
HEALTH SERVICES
Baywood Publishing Company, 120
Marine St., Farmingdale NY 11735,
USA. 4 issues for $25 p.a., $20
students, $42 inst.

ISIS-Women's International
Information and Communication
Service
Via Santa Maria dell'Anima 30,
Rome, Italy. 4 issues p.a. for $15, $25
inst. Also available: 'Women in
Development: a resource guide for
organization and action', $12;
'Women and New Technology'.

The JOURNAL OF COMMUNITY
COMMUNICATIONS
Village Design, P.O. Box 996,
Berkeley, CA 94701, USA. $9 for 4
issues, $15 institutions & Foreign.

The LEFT INDEX
511 Lincoln Street, Santa Cruz, CA
95060. 4 issues p.a. for $30 ind., $50
inst.

LITERATURE TEACHING
POLITICS
c/o Andrew Belsey, Dept. of
Philosophy, University College,
Cardiff CF1 1XL.

MEDICINE IN SOCIETY-
Quarterly Socialist Journal of Health
Studies.
16 St. John Street, London EC1. £6
p.a. individual, £7 inst., £9 overseas.

MIDNIGHT NOTES
Box 204, Jamaica Plain, MA 02130. 3

issues for $4. Special issues on the
anti-nuclear movement, the work/
energy crisis, space notes, the
computer state, political lemmings.
Back issues $2, or £1.50 from Radical
Science.

MODERNE ZEITEN (Socialist
Monthly)
Am Taubenfelde 30, 3000 Hanover 1,
W. Germany. 10 issues p.a. for DM
60.

MOTHER JONES-A Magazine
For The Rest of Us.
625 Third Street, San Francisco, CA
94107, USA. 10 issues p.a. for $18, $22
overseas.

MONTHLY REVIEW-An
Independent Socialist Magazine
62 West 14th St., New York, NY
10011, USA. 10 issues for $22
individual, $33 institutional p.a., $11
students, $18/$13 foreign.

MULTINATIONAL MONITOR
1346 Connecticut Ave., NW, Room
411, Washington D.C. 20036. 12 issues
p.a. for $18 individual, $25 non-profit
institutions, $35 business institutions,
foreign – add $8 air-mail.

NATURKAMPEN
c/o Politisk Revy, Vesterbrogade 31, 2
th., Dk-1620 Kbh., Denmark.

NEW GERMAN CRITIQUE- An
Interdisciplinary Journal of German
Studies
German Dept., Box 413, Univ. of
Milwaukee, WI 53201, USA. 3 issues
p.a. for $11 individual, $22 inst.,
foreign $1 extra.

OPEN ROAD-Anarcha-Feminist
Edition
Box 6135, Station G, Vancouver, B.C.,
Canada. Send 1 hour's pay for a sub.

OXFORD LITERARY REVIEW-A
Post-structuralist Journal
2 Marlborough Road, Oxford OX1
4LP, U.K. 2 issues p.a. for £5/$9, inst.
£6/$16.

PANDORE-Problems of Science,
Technology and Society
6 Blvd. St. Michel, 75006 Paris,
France.

PHILOSOPHY & SOCIAL
ACTION
M-120 Greater Kailash-1, New Delhi-
110 048, India. 4 issues p.a. for $15

ind., $35 inst.

POLITICS OF HEALTH
Newsletter
POHG, c/o BSSRS, Poland Street,
London W1. Send £2 + six sae's.

PRAXIS–A Journal of Radical
Perspectives on the Arts.
Dickson Arts Center, UCLA, Los
Angeles, CA 90024. 2 issues for $8; or
from Pluto Press, £5.

PROCESSED WORLD–The
Magazine With A Bad Attitude
55 Sutter Street, no. 829; San
Francisco, CA 94101. 4 issues for $10
ind., $15 inst. & Overseas.

**PSYCHOLOGY & SOCIAL
THEORY**
Triphammer Mall, Box 4387, Ithaca,
NY 14852. 2 issues p.a. for $12.50 ind.,
$25 inst.

RACE & CLASS–A Journal for
Black and Third World Liberation
Institute of Race Relations, 2–6 Leeke
Street, King's Cross Road, London
WC1X 9HS. 4 issues p.a. for £8/$16
individual, £12/$30 inst.

RACE TODAY–Voice of the Black
Community in Britain
165 Railton Road, London SE24. £6
ind., £14 inst.

RADICAL AMERICA
38 Union Square, Somerville, MA
02143, USA. 6 issues p.a. for $15
individual, $8 unemployed, foreign $3
extra; double for institutions.

RADICAL BOOKSELLER
265 Seven Sisters Road, London N4.
10 issues p.a. for £10 individual, £15
others.

**RADICAL COMMUNITY
MEDICINE** .
c/o Alex Scott-Samuel, 5 Lyndon
Drive, Liverpool L18 6HP, UK 4
issues p.a. for £6, £8 foreign (£11
airmail).

RADICAL HISTORY REVIEW
Mid-Atlantic Radical Historians'
Organization (MARHO), John Jay
College, 445 West 59th St., New York,
NY 10019, USA. 3 issues p.a. for $14
ind., $10 unemployed, $30 inst., plus
$4 extra abroad.

RADICAL PHILOSOPHY
c/o Ian Craib, Dept. Sociology, Univ.
of Essex, Colchester CO4 3SQ, UK

£3.25 for 3 issues p.a., £5 overseas.

RADICAL STATISTICS–Bulletin
of BSSRS Radical Statistics Group
9 Poland St., London W1. 3 issues
p.a. for £3, £5 inst., £1.50 for unwaged.

RED LETTERS–Communist Party
Literature Journal
16 St. John Street, London EC1.

**REVIEW OF AFRICAN
POLITICAL ECONOMY**
341 Glossop Road, Sheffield S10
2HP, UK. 3 issues p.a. for £6 in UK
& Africa, $13 elsewhere. No. 22 on
'Ideology, Class and Development',
£2.

**REVIEW OF RADICAL
POLITICAL ECONOMICS (URPE)**
155 W. 23rd St., 12th fl. New York,
NY 10011, USA. $4.50 per copy.

REVOLUON–Tijdschrift Over
Technologie, Natuurwetenschappen
en Kapitaal.
Postbus 1328, 6501 Nijmegen,
Holland. 4 issues for fl7.50 p.a.

The RIPENING OF TIME–
Theoretical Journal
P.O. Box 1103, 29 Mountjoy Square,
Dublin 1, Ireland.

SCHOOLING & CULTURE
ILEA Cockpit Arts Workshop,
Gateforth Street, London NW8. 3
issues p.a. for £5., £9 overseas.

SCIENCE & SOCIETY
Room 4331 John Jay College, CUNY,
445 West 59th St., New York, NY
10019, USA. 4 issues p.a. for $12,
foreign $19; $30 inst.

SCIENCE-FICTION STUDIES
Prof. Philmus, English Dept.
Concordia University, 7141
Sherbrooke St. West, Montreal,
Quebec, Canada H4B 1R6. 3 issues
p.a. for $12 individual/$19 inst., $10/
$16.50 in USA.

SCIENCE FOR PEOPLE–
Magazine of BSSRS
25 Horsell Rd, London N5. 4 Issues
p.a. for £4 individual, £10 inst. 10%
extra for foreign currency.

SCIENCE FOR THE PEOPLE–
Magazine of SESPA
897 Main Street, Cambridge, MA
02139, USA. 6 issues p.a. for $15, $24
inst. overseas, add $13 (airmail).

SE-SCIENZA ESPERIENZA

Via Valtellina 20, 20159 Milano, Italy. (This monthly magazine is a sequel to *Sapere*.)

SIGNS-Journal of Women in Culture and Society
University of Chicago Press, 5801 Ellis Avenue, Chicago, Illinois 60637, USA. 4 issues p.a. for $15., single copy $4.

SOCIALISM AND EDUCATION- Journal of the Socialist Education Association, 14 Branscombe St., London SE13 7AY. 3 issues for £1.50 p.a.; membership £4.

SOCIALIST HEALTH REVIEW 19 June Blossom Society, 60A Pali Road, Bandra (West), Bombay 400 050, India. 4 issues p.a. for $20.

SYGHRONA THEMATA c/o Giorgos Goulakos, Valaoritou 12, Athens 134, Greece.

TECHNOLOGY & CULTURE University of Chicago Press, 5801 Ellis Avenue, Chicago, Illinois 60637, USA.

TELOS-A Quarterly Journal of Radical Thought
Box 3111, St. Louis, MO 63130 USA. $22 for 4 issues p.a., $50 institutions, overseas, add 10%, cheques in US$ only.

TERMINAL 19/84-Centre d'Information et d'Initiative sur l'Informatisation
C.I.I.I., 1 Rue Keller, 75011 Paris, France.

TESTI E CONTESTI-Quaderni di Scienze, Storia e Societa.
CLUP, Piazza Leonardo da Vinci 32, Milano, Italy. 3 issues for Lit. 15,000, foreign 20,000.

UTUSAN KONSUMER Consumers Association Penang, No. 27 Kelawei Road, Pulau Pinang, Malaysia.

WECHSELWIRKUNG- Technik/Naturwissenschaft/ Gesellschaft
Gneisenaustr. 2, 1000 Berlin 61, W. Germany. 4 issues for 20 Dm.

WETENSCHAP EN SAMENLEVING VWO Stadhouderslaan 91, 3583 JG Utrecht, Holland. 10 issues for f35.

WIRE (Women's International Resource Exchange)
2700 Broadway, New York, NY 10025, USA. New free catalog includes reports on health care and Third World issues.

BACK ISSUES OF Radical Science Journal
STILL AVAILABLE

RSJ 5
BOB YOUNG: Science *is* Social Relations/PATRICK PARRINDER: Science and Social Consciousness in SF/DAVID TRIESMAN: The Institute of Psychiatry Sackings

RSJ 6/7 The Labour Process
LES LEVIDOW: A Marxist Critique of the IQ Debate/MIKE BARNETT: Technology and the Labour Process/BOB YOUNG: Getting Started on Lysenkoism/RSJ SUBGROUP: Marxism, Feminism and Psychoanalysis/ LES LEVIDOW: Grunwick as Technology and Class Struggle

RSJ 8
DAVID DICKSON: Science and Political Hegemony in the 17th Century/ WENDY HOLLWAY: Ideology and Medical Abortion/PHILIP BOYS: Detente, Genetics, and Social Theory

RSJ 9 Medicine
KARL FIGLIO: Sinister Medicine?/GIANNA POMATA: Seveso – Safety in Numbers?/LES LEVIDOW: Three Mile Island/Critical Bibliography on Medicine

RSJ 10 Third World
DAVID DICKSON: Science and Technology, North and South: Multinational Management for Underdevelopment/RAPHAEL KAPLINSKY: Microelectronics and the Third World/LES LEVIDOW: Notes on Development/BRIAN MARTIN: The Goal of Self-Managed Science

RSJ 11
RSJ COLLECTIVE: Science, Technology, Medicine and the Socialist Movement/JONATHAN REE: The Anti-Althusser Bandwagon/PAM LINN: Designer or Drone?/MAUREEN McNEIL: Braverman Revisited

RSJ 12 Medicalisation
INTRODUCTION: Unnatural Childbirth?/JANET JENNINGS: Who Controls Childbirth?/SHELLEY DAY: Is Obstetric Technology Depressing?/EVAN STARK: What is Medicine?

RSJ 13 Scientism in the Left
STEVE SMITH: Taylorism Rules OK? Bolshevism, Taylorism and the Technical Intelligentsia/LES LEVIDOW: We Won't Be Fooled Again? Economic Planning and Left Strategies/DOUG KELLNER: Science and Method in Marx's Capital/JOE CROCKER: Sociobiology: The Capitalist Synthesis/TIM PUTNAM: Proletarian Science?

If ordering direct with payment in foreign currency, add equivalent of 60p to cover bank charges and use current exchange rates.
Radical Science Journal, 26 Freegrove Road, London N7 9RQ

RADICAL SCIENCE provides a forum for extended analyses of the ideology and practice of science, technology and medicine from a radical political perspective. Most contributors have attempted to re-examine past Marxist views and to develop a Marxist critique of the role of scientism in the Left.
We welcome suggestions for thematic numbers emphasising particular topics around the critique of power exercised through expertise. We would like especially to develop a positive programme for the role of oppositional knowledge.

Radical Science Collective
This serial is edited and produced by a collective whose members are Gavin Browning, Joe Crocker, Karl Figlio, Mike Hales, Chris Knee, Les Levidow, Pam Linn, Maureen McNeil, Barry Richards, Tony Solomonides, Margot Waddell and Bob Young.

Overseas Contacts
ITALY: Gianna Pomata, Via Bertiera 7/2, Bologna.
FRANCE: John Stewart, 27 rue de Montreuil, Paris 75011. Tel. 356.10.71.
David Dickson, Le Billehau, St Aubin, 91190 Gif-su-yvette.
AUSTRALIA: Brian Martin, Maths. Faculties, ANU, PO Box 4, Canberra ACT 2600 Tel. (062) 494445, home tel. (062) 485426

Editorial Contributions
Ideally, articles should be less than 10,000 words and typed double-spaced. A number of copies would help: it is our policy that all articles should be read by as many members of the collective as possible. This usually takes some time so please bear with us – but remind us when your patience runs out.

Subscriptions to RADICAL SCIENCE
The subscription for three numbers is £14.00 post paid, libraries and institutions £18.00. Numbers appear at irregular intervals, so the subscription is not annual: it covers three consecutive numbers. For details of back numbers see inside rear cover. Those who support us in our project are invited to add a donation to their subscription. Subscription to *Free Associations* (published quarterly): £20/$25 per annum for individuals; £35/$45 for institutions. Single copies: £5.50/$6.50. Please add the equivalent of £0.60 to foreign cheques to cover bank charges involved. Subscriptions, donations, enquiries and articles should be sent to:
   Free Association Books, 26 Freegrove Road, London N7 9RQ. Tel. (01) 609 5646.

Distribution
UK: Turnaround, 25 Horsell Road, London N5. Tel. (01) 609 7836.
S&N, 48a Hamilton Place, Edinburgh EH3 5AX. Tel. (031) 225 4590.
18 Granby Row, Manchester M1 3GE. Tel. (061) 228 3903.
USA: B.deBoer, 113 East Center Street, Nutley, NJ 07119. Tel. (201) 667 9300.
Carrier Pigeon, 40 Plympton St., Boston, MA 02118. Tel. (617) 542 5679.
CANADA: DEC, 427 Bloor St West, Toronto, Ontario M5S 1X7. Tel. (416) 964 6560.
AUSTRALIA: Astam Books, 250 Abercrombie Street, Chippendale, NSW 2008. Tel. Sydney 698 4080.

Copyright